D0858407

CLONALITY

CLONALITY

The Genetics, Ecology, and Evolution of Sexual Abstinence in Vertebrate Animals

JOHN C. AVISE

WITH ANIMAL ILLUSTRATIONS
BY TRUDY NICHOLSON

OXFORD
UNIVERSITY PRESS
2008

OXFORD
UNIVERSITY PRESS

Oxford University Press, Inc., publishes works that further
Oxford University's objective of excellence
in research, scholarship, and education.

Oxford New York
Auckland Cape Town Dar es Salaam Hong Kong Karachi
Kuala Lumpur Madrid Melbourne Mexico City Nairobi
New Delhi Shanghai Taipei Toronto

With offices in
Argentina Austria Brazil Chile Czech Republic France Greece
Guatemala Hungary Italy Japan Poland Portugal Singapore
South Korea Switzerland Thailand Turkey Ukraine Vietnam

Published by Oxford University Press, Inc.
198 Madison Avenue, New York, New York 10016

www.oup.com

Library of Congress Cataloging-in-Publication Data
Avise, John C.
Clonality : the genetics, ecology, and evolution of sexual abstinence in vertebrate
animals / John C. Avise ; with animal illustrations by Trudy Nicholson.
p. cm.
Includes bibliographical references and index.
ISBN 978-0-19-536967-0
1. Vertebrates—Reproduction. 2. Reproduction, Asexual. I. Title.
QP251.A95 2008
571.3'16 dc22 2008004049

Cover image: Frog in the genus *Rana*. Frogs have been important in scientific
studies of clonal reproduction. One frog species was the first vertebrate
artificially cloned in the laboratory, and in nature, some frogs have a breeding
system with elements of both clonality and sexuality.

9 8 7 6 5 4 3 2 1

Printed in the United States of America
on acid-free paper

PREFACE

The word "cloning" sometimes conjures images of mad scientists in white lab coats, all working diligently to produce legions of genetically identical humans in a futuristic Orwellian society. These nightmarish scenarios are technically feasible in the modern era of genetic engineering (though hopefully no one will actually pursue them). But cloning also has far more benign connotations. In nature it is a standard asexual reproductive operation of many plants and animals, including various vertebrates—creatures with backbones, such as particular species of fish, amphibians, reptiles, and mammals—that reproduce, in effect, without the benefit of standard sexuality. In this book, we will revel in the diversity of clonal modes in vertebrate animals, and we will ponder the ecological and evolutionary ramifications of asexual and quasi-sexual reproductive systems associated with clonality. Clonal reproduction also offers an intriguing vantage point for interpreting the biological significance of sexual reproduction, which has long been one of evolution's greatest enigmas.

The reproductive lifestyles of clonal creatures can seem so outlandish as to nearly defy belief. For example, several wild species of reptiles, amphibians, and fish consist only of females, each of whom reproduces by making near-perfect genetic replicas of herself. In some of these unisexual taxa, genuine virgin births take place. In others, each female requires the sexual services of a male from a foreign species. The "sexually parasitized" male may play only a stimulatory role, but he may in some cases make a genetic contribution to his daughters. This latter situation seems even more bizarre when we consider that each daughter may manipulate her father's chromosomes in ways that prevent him from becoming a genetic grandfather of her offspring!

In many other animal species, an individual can be both a male *and* a female during its lifetime. In one such species of fish, each hermaphroditic individual normally fertilizes itself. When this selfing process continues generation after generation, highly inbred lines emerge in which different fish are near-exact genetic replicas of one another. Thus the members of each inbred lineage in effect are clonemates. And lest you think vertebrate clonality is confined to "lowly" fish, reptiles, and amphibians, consider the nine-banded armadillo. In this peculiar little mammal, each pregnancy yields quadruplet pups that are clonemates of each other yet genetically different from both parents. By a similar genetic process, humans also occasionally produce clonemates (e.g., identical twins).

Clonal operations also occur at other levels of biological organization. DNA replication is basically a clonal process, as is cellular proliferation during the development of a multicellular organism. Furthermore, some segments of DNA in all animals are transmitted solely via females and others solely via males, without the genetic shuffling that accompanies sexual inheritance in most other genes. For such uniparentally inherited DNA, clonal replication extends indefinitely across successive animal generations, in sexual species as well as those that reproduce asexually.

With the invention of genetic engineering technologies in recent years, clonality is no longer solely under nature's purview; the phenomenon now occurs in research labs. Borrowing molecular tools from nature (and inventing some new ones of their own), scientists have learned how to generate "artificial" clones of particular genes and cells and also of whole animals.

Periodically during my 40-year career as an evolutionary geneticist, I have conducted research on a medley of asexual genetic systems and clonally reproducing creatures. Nature's diverse routes to clonality provide interesting and informative departures from the familiar norms of genetic recombination that accompany sexual reproduction. In this book, I describe the full panoply of asexual operations known in vertebrate animals, sometimes integrating cellular and organismal phenomena that typically have not been unified under the conceptual umbrella of clonality. I discuss the genetic mechanisms underlying various forms of clonality, as well as the ecological and evolutionary ramifications of different classes of asexual and quasi-asexual reproduction. I also describe the many vertebrate species that display each type of clonality. A central theme, based on empirical evidence, is that clonal reproduction can be a highly successful evolutionary strategy in vertebrate animals, but only as a short-term opportunistic tactic. Beyond conveying this seminal message, this book is meant to be entertaining as well as educational for a wide audience of biologists and natural historians.

Chapters in this book are organized into four sections. Part I sets the stage by describing intraindividual aspects of clonality and addressing the fundamental distinctions between clonal and sexual reproduction. Part II examines unisexual (all-female) vertebrate taxa that display sperm-independent (parthenogenetic) and sperm-dependent (gynogenetic, hybridogenetic, and kleptogenetic) forms of clonality. Part III addresses two other forms of vertebrate clonality in nature, via polyembryony and extreme incest. Part IV considers various forms of vertebrate clonality engineered by humans. The book also contains a glossary and an extensive list of cited references to the primary scientific literature.

I am grateful to Trudy Nicholson for providing the beautiful line drawings of various clonal animals that grace this volume (figs. 3.3–3.16, 4.3–4.13, 5.1, 5.2,

6.2, 6.3, 7.3, 7.6, 7.7, and the frontispiece). Trudy and I have produced several books together, and it is always a joy for me to work with this conscientious and gifted artist. Thanks go to Dr. Jim Bogart, to the current members of my laboratory (Felipe Barreto, Rosemary Byrne, Vimoksalehi Lukoschek, and Andrey Tatarenkov), and to several anonymous reviewers for helpful comments on early drafts of this book. I want to dedicate this work to my mother and to the loving memory of my father. My parents raised me with passions for natural history and for intellectual inquiry that have always been central parts of my life. I also want to dedicate this book to my devoted wife and daughter, who to my continual amazement are unfailingly supportive of my efforts.

CONTENTS

CLONALITY

PART I

Background: Like Begets Like

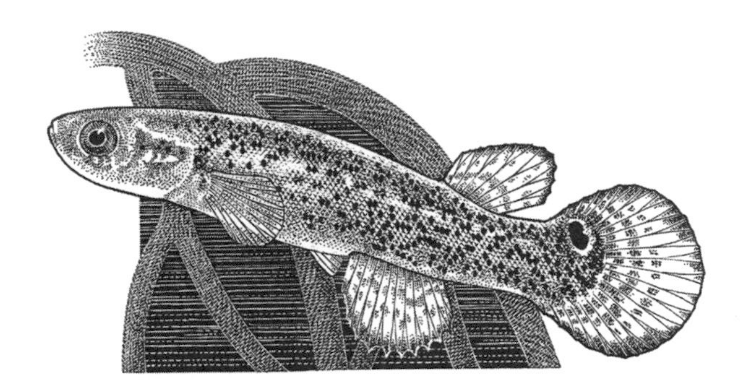

A clone can be defined as a genetic copy (or all such alike descendants) of a previously existing biological entity. Such an entity can be interpreted broadly to include a particular stretch of DNA (a locus), an ensemble of physically linked loci (such as the genes comprising mitochondrial DNA), a genome (the entire suite of DNA) of a somatic cell, or the full genetic constitution of a multicellular organism. In vertebrates, clonal replication occurs universally at the first three of these levels, but clonality at the whole-animal level is exceptional. This book focuses primarily on organismal clones, with the chapters in part I setting the empirical and conceptual backdrop. Chapter 1 introduces intraindividual aspects of clonality that are necessary for understanding how nature sometimes produces whole-organism clones. Chapter 2 then introduces some of the ecological and evolutionary quandaries posed by the striking contrast between clonal and sexual tactics of vertebrate reproduction.

CHAPTER ONE

Clonality within the Individual

All forms of clonal reproduction begin with the faithful replication of genetic material. When James Watson and Francis Crick announced their discovery of the double-helical structure of deoxyribonucleic acid in 1953, they immediately recognized DNA's dual roles in storing vast amounts of hereditary information *and* in providing an ideal template for self-replication. Four types of nucleotides—containing the organic bases adenine, guanine, thymine, or cytosine—are the biochemical building blocks of DNA, and they can be thought of as letters that compose the words, sentences, and paragraphs of any genomic text. Each DNA molecule is composed of two complementary strands of nucleotides, intertwined like mating cobras. Along this double helix, each adenine in one strand is paired (by hydrogen bonds) to a thymine in the other strand, and each guanine is paired to a cytosine. Thus the order of nucleotides in one strand predicts the nucleotide order of its wedded partner. The two strands are structurally redundant but non-identical, like positive and negative photographs of a physical object.

Clonality at the Gene Level: DNA Replication

Watson and Crick surmised that during gene replication, the two biochemical strands of DNA first unzip, after which each strand can serve as a template for assembling (from a pool of unbound nucleotides within the cell) its complementary twin, thereby generating two identical double helices where formerly there was one (fig. 1.1). Watson and Crick's classic paper in *Nature* focused on DNA's molecular structure rather than its mode of replication, but in the concluding paragraph the authors coyly wrote, "It has not escaped our notice that the specific pairing we have postulated immediately suggests a possible copying mechanism" (1953, p. 737). In a follow-up paper also published in 1953, Watson and Crick elaborated their hypothesis for DNA replication: "We imagine that prior to duplication the hydrogen bonds are broken, and the two chains unwind and separate. Each chain then acts as a template for the formation onto itself of a new companion chain.... Moreover, the sequence of the pairs of bases will have been duplicated exactly" (p. 966).

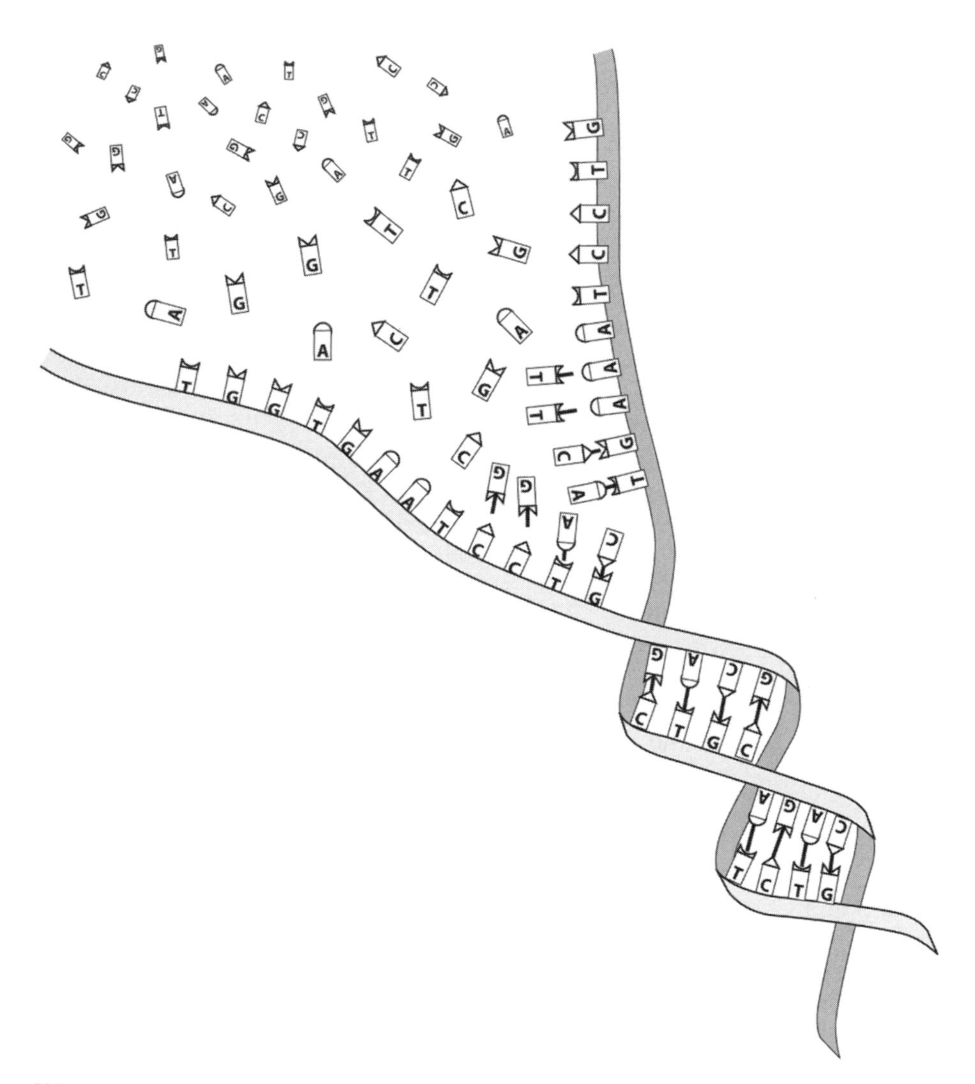

FIGURE 1.1 When DNA replicates, the double helix unzips and new nucleotides are incorporated.

In 1958, Matthew Meselson and Franklin Stahl experimentally confirmed Watson and Crick's prediction that each of the two daughter molecules arising from a DNA replication event consists of one old strand from the parent molecule and one newly synthesized strand from a cell's pool of free nucleotides. Working with colonies of *Escherichia coli,* Meselson and Stahl grew the bacteria for several generations in a culture medium containing a heavy isotope of nitrogen (^{15}N), which the cells incorporated into both strands of their DNA. The bacteria were then transferred to a medium with normal nitrogen (^{14}N) and allowed to reproduce. After one round of DNA replication on the new medium, one strand in each double helix carried only ^{15}N and the other carried only ^{14}N. After two rounds of DNA replication, 50% of the DNA molecules consisted of newly generated strands with ^{14}N only, and the other 50% consisted of one old strand (solely with ^{15}N) and one new strand (solely with ^{14}N). These outcomes were exactly as Watson and Crick had predicted, and they eliminated two competing models for how DNA might otherwise replicate (fig. 1.2). Subsequent studies showed that the basic process of DNA replication in multicellular organisms, vertebrates included, is similar to that in *E. coli.*

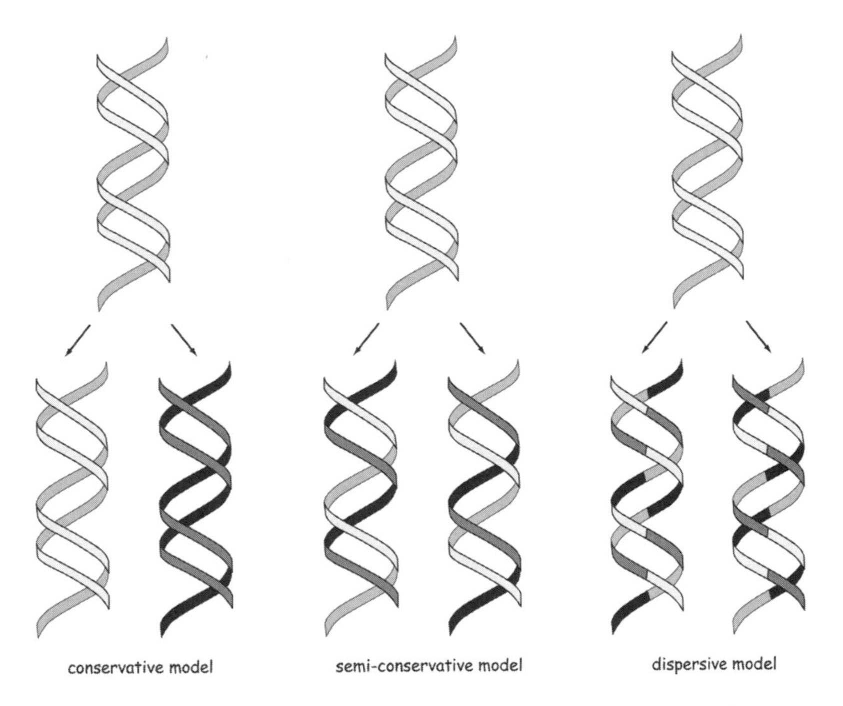

FIGURE 1.2 Three earlier hypotheses for how DNA's double helix might replicate. The semiconservative model proved to be correct. In each bottom diagram, the heaviest shading indicates newly synthesized strands of DNA.

In biology textbooks, this mode of DNA replication is termed semiconservative. For pedagogic purposes it is usually contrasted with the experimentally falsified conservative model for DNA replication (fig. 1.2), in which each double helix remains intact while generating a duplicate copy. Ironically, however, in the context of discussing clonal reproduction, use of the word "semiconservative" to describe DNA replication could be misconstrued. The standard replication of genetic material in proliferating cells yields exact copies (barring mutation; see box 1.1) of DNA templates, so the genetic information is faithfully preserved (conserved). In this important descriptive sense, the semiconservative nature of DNA replication is really highly conservative.

Clonality at the Genomic Level: Mitosis

Inside the nucleus of each animal cell, DNA molecules and proteins form a scaffold of chromatin fibers that in turn are arranged into threadlike structures known as chromosomes. In diploid cells that comprise most of an individual's body or soma, these chromosomes come in homologous pairs (homologues), one member tracing back to a haploid sperm of the sire and the other to the haploid egg of the dam. In a human being, for example, each somatic cell normally contains 22 pairs of autosomes plus two sex chromosomes (XX in females, XY in males), for a total of 46 chromosomes (23 from dad, 23 from mom). Each haploid chromosome set contains DNA sequences totaling more than three billion nucleotide pairs, all ordered into approximately 30,000 functional genes (those that specify ribonucleic acids and proteins), plus vastly longer stretches of noncoding DNA whose functional significance, if any, often remains unknown. Comparable statements apply to all other sexually reproducing vertebrates, with only technical amendments for species differences in chromosome numbers, genomic sizes and arrangements, or modes of sex determination.

In sexually reproducing species, each individual begins life as a single diploid cell borne of a genetic union between sperm and egg. That fertilized egg or zygote then divides into two cells, those two into four, four into eight, and so on to great numbers—about 100 trillion in an adult human. With respect to morphology, biochemistry, and physiology, these somatic cells are highly diverse, as illustrated by the profound differences between myoblasts (muscle-producing cells), neurons (nerve cells), hair-follicle cells, or those of the liver or kidney. With respect to DNA sequences, however, all somatic cells of an individual are identical, barring de novo mutations. Thus the differences in form and function among an individual's somatic cells and organs are due to altered patterns of gene expression rather than to sequence changes in the genetic blueprints themselves. Precisely how genes are regulated (turned on, off, or modulated) during

BOX 1.1 De Novo Mutations Are Inherent Imperfections in the DNA Replication Process

Especially during DNA replication, new mutations routinely interconvert or otherwise alter small fractions of nucleotides—much as a scribe might introduce occasional spelling errors when copying a written text. Most of these biochemical mistakes are quickly recognized and mended by legions of DNA-repair enzymes inside cells, but this molecular copyediting is not perfect, and at least a few errors almost inevitably escape detection.

Consider, for example, the process of DNA replication in a typical mammalian genome that is about three billion base pairs (bp) in composite length. Empirically, de novo mutations accumulate at a rate of approximately 10^{-9} per nucleotide site per animal generation. So about three nucleotides in a mammalian genome are likely to experience and retain a new mutation during any standard clonal operation that otherwise would yield a perfect genetic replica of the original. Thus, in our mammalian example, a first-generation clone might only be 99.999999999% identical to its predecessor.

Mutational processes mean that "100% pure" clones seldom exist, particularly at the levels of entire genomes or multicellular organisms. For this reason, recent-origin mutations are normally excepted when referring to clonemates or to clonal lineages, members of which are nearly, but not necessarily fully, identical in genetic composition (e.g., Schartl et al., 1991; Malysheva et al., 2007). The genuine essence of any form of cloning is the faithful reproduction of genetic material *in the absence of genetic recombination,* that is, without sex (or other means) by which DNA molecules from separate biological sources are otherwise mixed and shuffled during nonclonal (sexual) reproduction.

De novo mutations sometimes cause genetic disabilities and understandably have a bad reputation. It is worth remembering, however, that mutations are the ultimate source of genetic variety, including variation that is shuffled in each round of sexual reproduction into novel and sometimes adaptive combinations. If mutations were somehow to cease entirely, evolution and life itself would eventually expire.

postzygotic development is the key to ontogeny, and a main research focus in the field of developmental biology.

For current purposes, the important point is that all of an individual's somatic cells are clonally derived. The universal cellular process that achieves this outcome is mitosis. During each mitotic event, all of a cell's chromosomes and their

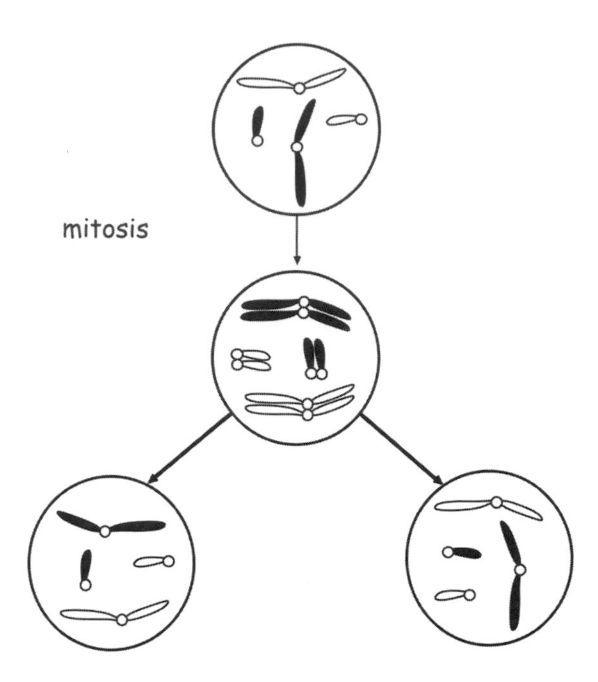

FIGURE 1.3 Simplified diagram of mitosis, showing clonal outcomes of the process.

constituent DNAs are faithfully copied and distributed equally to two daughter cells (fig. 1.3). One round of genomic replication transpires per cell division, and one of the duplicate genomes is distributed to each derived cell. Hence daughter cells that emerge from each mitotic event—and, by extension, all diploid cells of an individual—are miniature genetic clonemates with respect to their nuclear DNA.

Clonality in the Cellular Cytoplasm: Mitochondrial DNA

Mitochondria—tiny cytoplasmic organelles that help to generate and store a cell's chemical energy—are home to another important category of DNA. Each mitochondrial DNA, or mtDNA, molecule (fig. 1.4) is a closed circle about 16,000 nucleotide pairs long, and several hundred copies typically inhabit each somatic cell (in sharp contrast to nuclear loci, which occur in only two copies per diploid cell) (Satoh and Kuroiwa, 1991). Nonetheless, all mtDNAs within an individual are essentially identical in nucleotide sequence because the molecules proliferate,

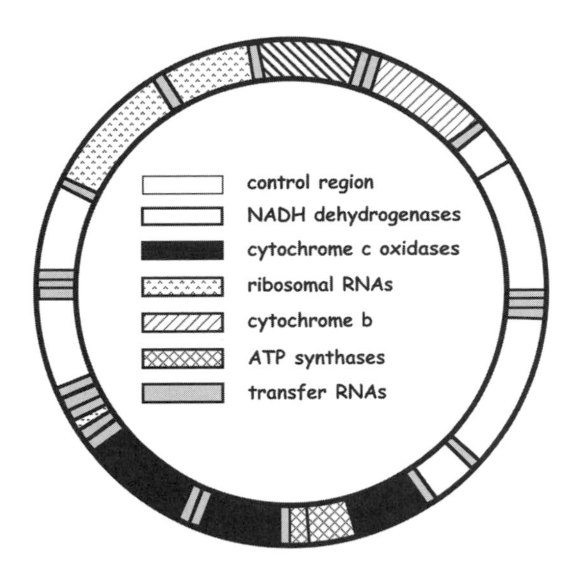

FIGURE 1.4 Animal mitochondrial DNA, or mtDNA. Each molecule is typically composed of two loci specifying ribosomal RNAs, 22 loci specifying transfer RNAs, and 13 genes that encode protein subunits. Although mtDNA thus consists of about 37 functional genes, in effect the whole molecule is one evolutionary "supergene" because it transfers across the generations as a single unit, without undergoing genetic recombination.

though at a much faster pace than cell divisions, by otherwise standard modes of clonal DNA replication.

Furthermore, mtDNA differs from nuclear DNA in being maternally transmitted from parent to progeny, typically without appreciable genetic contribution from the father (box 1.2). This happens in large part because, when an organism is conceived during the fertilization process, the zygote's cytoplasm (and hence most of its mtDNA) comes primarily from the dam's egg rather than the sire's sperm. The paucity or absence of paternally derived mtDNA in offspring has a major genetic ramification: paternal and maternal mtDNA genes seldom, if ever, recombine, even during sexual reproduction. In other words, mtDNA molecules are transmitted clonally not only during successive cell generations within an individual, but also from one animal generation to the next. Absent the cellular and genetic processes that routinely shuffle nuclear DNA during sexual reproduction, mtDNA in effect is celibate and asexual across as well as within animal generations.

The clonal mode of maternal inheritance for mtDNA is quite analogous to the paternal inheritance of human surnames (Avise, 1989). In many human populations, intact surnames (unaltered by maternal input) are transmitted through paternal lineages, much as intact mtDNA molecules (unaltered by paternal input) are transferred through maternal lineages. Of course, surname transmission is a recent social convention unique to particular human cultures,

BOX 1.2 Clonal Maternal Inheritance of Animal mtDNA

Interestingly, an all-female species of quasi-clonal fish (which will be described further in chapter 4) provided an unusually critical experimental test of the possibility of *paternal leakage* of mtDNA into an otherwise matrilineal pedigree. In this unisexual species, *Poeciliopsis monacha-lucida*, females mate in each generation to males of a congeneric sexual species such that each daughter carries two distinct nuclear genomes—one of maternal and the other of paternal origin. Do the daughters carry some paternally derived mtDNA as well?

In effect, the *Po. monacha-lucida* females are engaged in perpetual backcrosses to males of a foreign species. Thus, if even a tiny fraction of paternal mtDNA were to leak into the unisexual species in each generation (via the mtDNA-carrying sperm that fertilize the females' eggs), over time the distinct paternal mtDNA could build up to readily detectable levels and would "stand out like a sore thumb" against the backdrop of maternally derived mtDNA in the unisexual line. Robert Vrijenhoek and I (1987) examined *Po. monacha-lucida* individuals for evidence of male-derived mtDNA, but no such genetic signature of paternal leakage was found. (Our studies were conducted under laboratory assay conditions that could have detected a rate of paternal leakage as few as one in 25,000 mtDNA molecules per generation). Results supported then-conventional wisdom that mtDNA is strictly maternally inherited in vertebrate animals.

Subsequent genetic studies have uncovered scattered instances of low-level paternal leakage in various other animal taxa (e.g., Kondo et al., 1990; Avise, 1991; Gyllensten et al., 1991). Perhaps this is not too surprising, because each sperm cell does carry a small number (i.e., dozens or scores) of mtDNA molecules (compared to many thousands in a mature oocyte), and some of these may occasionally colonize a zygote's cytoplasm during the fertilization process. Still, the overwhelming conclusion remains that mtDNA is transmitted predominantly, if not exclusively, though matrilines in probably all vertebrate species.

Another point is that various recombinational mechanisms do exist in animal mtDNA (see Birky, 2001; McVean, 2001; Rokas et al., 2003), including vertebrate mtDNA (Tatarenkov and Avise, 2007), so in this sense the molecule is not strictly clonal. However, the vast majority of recombination events presumably involve genetically identical or nearly identical mtDNA molecules within a female lineage, rather than foreign mtDNA molecules introduced by sperm. Vertebrate mtDNA therefore remains, in most cases, essentially clonal.

whereas mtDNA transmission is an ancient genetic convention almost universal to multicellular forms of life. Because mtDNA is transmitted clonally from mothers to progeny of both sexes, mtDNA is sometimes referred to as a non-Mendelian genetic system. As described later in this book, it does indeed violate the two fundamental laws of heredity that Gregor Mendel deduced for nuclear genes more than 150 years ago.

Genetic alterations do nevertheless arise and gradually accumulate in mtDNA molecules, due mostly to de novo mutations that typically involve simple substitutions of one nucleotide for another. Indeed, mtDNA nucleotide sequences in many vertebrate species evolve about 5–10 times faster than typical nuclear DNA sequences. This fast pace of evolution is probably due to at least three contributing factors (Lynch et al., 2006): the elevated presence in mitochondria of oxygen radical molecules that tend to produce mutations; the fact that mtDNA molecules are continuously replicated, with high nucleotide-misincorporation rates, even within nondividing cells; and the likelihood that mitochondria have less efficient mechanisms of DNA repair than those of the cell's nucleus. As described in Part II, a rapid pace of evolution helps to make mtDNA a superb molecular marker for deducing the evolutionary origins and ages of clonal vertebrate lineages.

Clonality in the Sex Chromosomes

Most chromosomes in the nucleus of a diploid vertebrate come in matched pairs called autosomes, which during meiosis routinely pair up and exchange parts (see chapter 2). Sex chromosomes, by contrast, come as an unmatched pair. The Y chromosome of mammals, for example, is normally much smaller than the X and carries many fewer genes. Because major portions of the X and Y are dissimilar, they fail to align and exchange DNA sequences during meiosis. Furthermore, unlike the case for X chromosomes, two Y chromosomes almost never co-occur within an individual, because XX is the "homogametic" sex in mammals and XY is the "heterogametic" sex. So the Y chromosome in effect is transmitted clonally across animal generations, through patrilines, in somewhat analogous fashion to how mtDNA is transmitted clonally along matrilines. The X chromosomes, by contrast, are more like pairs of autosomes in the sense that they do recombine with one another during meiosis, albeit only in females.

A similar but mirror-image situation applies in birds, which have a ZW system of sex chromosomes in which females are the heterogametic sex (ZW) and males are homogametic (ZZ). Thus the always-unpaired W chromosomes are transmitted through matrilines and are effectively clonal, whereas the Z chromosomes can recombine during meiosis in males. In other vertebrates such as fish, some species have an XY system, others have a ZW system, and some species lack distinguishable sex chromosomes altogether (Mank et al., 2006).

It is rather ironic that some of the "sex" chromosomes (Y and W) are in effect asexually transmitted. They nicely illustrate the point that clonal replication—across animal generations as well as within the soma—can apply not only to cytoplasmic genomes but also to particular subsets of the nuclear genome.

BOX 1.3 Gene Conversion, Mitotic Recombination, and Epigenetic Drift

The intended meaning of "genetic identity" or clonality, as used in this book, requires clarification. Apart from de novo DNA mutation per se (box 1.1), at least three categories of molecular process can make clonemates (cells, or entire organisms) slightly less than 100% genetically identical. The first is "gene conversion" (Ohta, 2000), a cellular operation by which particular genes of a multigene family occasionally convert other members of the family to their own DNA sequence. Many loci are present in two or more copies (sometimes thousands) within a genome, sometimes tandemly aligned and sometimes dispersed across the chromosomes. The gene conversions that these loci may undergo involve recombination-like processes, but such events are not to be confused with the traditional sex-mediated recombination, via meiosis and syngamy, that applies to nearly all nuclear genes (see chapter 2). Gene conversion events in effect shuffle multilocus nuclear DNA sequences to some degree; they can thereby generate detectable genetic diversity, even among otherwise clonemate cells or individuals (Gorokhova et al., 2002). Gene conversion events are also suspected for some duplicated gene regions of animal mtDNA (Tatarenkov and Avise, 2007).

"Mitotic recombination"—the reciprocal exchange of homologous chromosome segments via atypical crossover events during mitosis—is another cytological process that can introduce occasional genetic variation into proliferating cells that are otherwise clonal (e.g., Shao et al., 1999). This phenomenon is relatively rare, however, and thus is not a serious complication in most clonality discussions.

A third process that in effect can produce genetic differences among clonemate cells or organisms is "epigenetic drift" (Martin, 2005). During an animal's development, many genes are biochemically modified or otherwise differentially regulated (activated or silenced) in different somatic cells and tissues. Such alterations in gene expression are extremely important in ontogeny (Bird, 2007), and they can also yield small but detectable epigenetic differences between otherwise identical monozygotic twins (Fraga et al., 2005). For purposes of this book, however, epigenetic variation, like gene conversion and mitotic recombination, can henceforth be safely neglected.

Clonality among the Soma

During ontogeny, the processes of DNA replication (mitochondrial and nuclear) and mitotic cell division ensure that the trillions of somatic cells comprising any vertebrate animal are genotypically identical, in essence (but see box 1.3). Accordingly, an individual can be viewed as a huge symbiotic colony of asexually derived clonemate cells, all normally working together for the common good. That collective "good" is the successful transmission of gene copies to subsequent generations, beyond the lifetime of the mortal individual. All of an animal's somatic cells have a shared and vested ultimate interest in this collaborative reproductive endeavor.

Indeed, multicellularity itself would normally be untenable if the collaborating cells were not clonemates. Imagine a hypothetical world in which each multicellular animal begins life as an amalgam of genetically unrelated cells. That situation would seldom be sustainable (but see Moran, 2007, for possible exceptions) because rampant intercellular disputes and genetic conflicts of interest would inevitably arise, precluding the evolutionary maintenance of close functional collaborations required of any reproducing biological entity. At the other end of the hypothetical spectrum, imagine a biological world in which each multicellular organism *is* comprised of clonemate cells, but each daughter individual also arises asexually and is thus a multicellular clone of her parent. As we shall see later, this situation actually does characterize many vertebrate taxa. But before turning to such organisms in chapter 3, we must first consider the antithesis of clonality: sexuality, the subject of chapter 2.

CHAPTER TWO

Sexuality: The Antithesis of Clonality

Sexual reproduction or amphimixis—genetic recombination via gamete formation and union—provides a dramatic departure from the clonal mode of genomic replication that accompanies the mitotic propagation of somatic cells within an individual (see chapter 1), or, in asexual species, the clonal propagation of individuals across generations (see chapters 3 and 4). With regard to nuclear DNA, the genotypic identity of somatic cells within an individual stands in striking contrast to the genotypic dissimilarity between somatic cells of a parent and its sexually produced offspring. By combining its own genes with those of another individual, a sexual dam or sire in effect dilutes its personal genetic contribution to each offspring by 50% (compared to a hypothetical standard in which it had produced a clonal copy of itself). Another way of viewing this is to note that with a fixed level of investment in offspring, a sexual female can produce only half as many daughters as would her clonally reproducing counterpart. This "twofold cost of sex"—brought eloquently to the attention of evolutionary biologists by John Maynard Smith in 1971—is one reason why the prevalence of sexual reproduction has long been an evolutionary enigma.

Sex at the Interchromosomal Level: Meiosis, Syngamy, and Mendel's Laws

Gregor Mendel was the first person to deduce what happens at the genetic level during sexual reproduction. This Austrian monk conducted his research in the mid-1800s, 50 years before the word "gene" was introduced and a full century before DNA was confirmed to be life's hereditary material. Mendel had no powerful microscopes or other fancy investigative tools of molecular biology; instead, he gained his remarkable insights by carefully monitoring crosses between true-breeding strains of pea plant with contrasting features, such as tall versus short stems or round versus wrinkled seeds. By carefully tallying how many progeny of each type resulted from each cross, Mendel unearthed two cornerstone principles of nuclear gene heredity that have proved to be near-universal hallmarks of sexual reproduction.

Mendel's first principle—the law of segregation—applies to nuclear genes considered individually. It refers to the fact that during egg production by a female, or sperm (or pollen) production by a male, each haploid gamete receives one copy (i.e., allele) of each nuclear gene, drawn at random, from the parent's diploid genotype. Recall that a diploid cell carries pairs of homologous chromosomes, and thus thousands of pairs of homologous genes (loci). At each locus, any parent's gametic pool (its ensemble of eggs or sperm) contains both alleles, but each haploid egg or sperm cell carries only one. When two gametes unite to form a zygote during the fertilization process, the diploid cellular condition is thereby restored.

Mendel's second principle—the law of independent assortment—refers to the fact that the process of allelic segregation during gametogenesis (gamete formation) occurs independently across unlinked loci (those on different chromosomes). In other words, the random segregations of alleles at gene A and at gene B are statistically uncoupled such that the dilocus (and by extension multilocus) allelic combinations distributed to gametes are also randomized.

Although Mendel could not have known it at the time, the two hereditary principles that he uncovered reflect the behaviors of nuclear chromosomes during meiosis—the cellular process underlying gametogenesis (fig. 2.1). During each meiotic dance on the cellular stage in an ovary or testis, nuclear chromosomes line up with their homologous partners, exchange parts, and then separate, the members of each pair finally segregating from one another and assorting independently of other such pairs. Overall, this finely tuned process entails one round of genomic duplication and two successive rounds of cell division. As a result, from each meiotic event that began with a diploid precursor cell (an oogonium in the ovary or a spermatogonium in the testis), four haploid cells emerge, one or more of which will mature into functional gametes.

The choreography of meiosis is sophisticated and the chromosomal dancers are skilled. So there is considerable irony in the fact that meiosis seems geared to produce gametes with "haphazard" multilocus combinations of the parent's alleles. When a sperm and an egg later join in syngamy to form a diploid zygote, the full process of genetic recombination via sex is finally consummated (fig. 2.1).

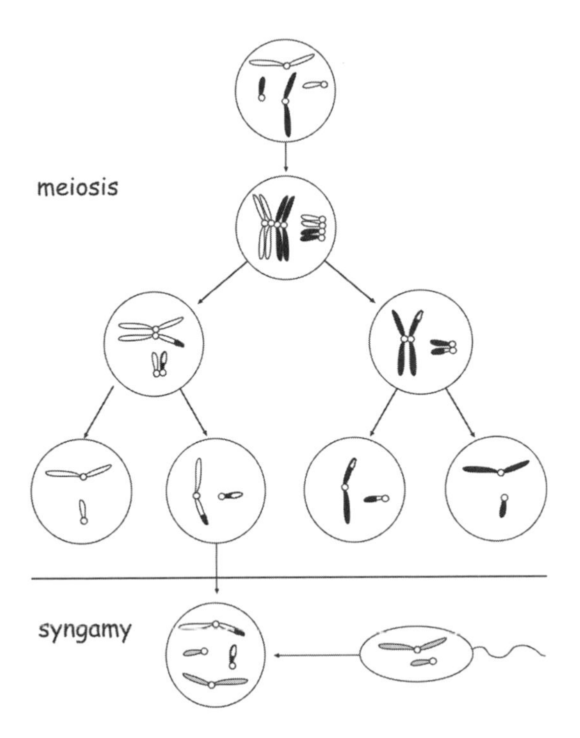

FIGURE 2.1 Simplified diagram of meiosis and syngamy, showing how genes are mixed during sexual reproduction (see also fig. 2.2).

Sex at the Intrachromosomal Level: DNA Recombination via Crossing Over

Although Mendel's law of independent assortment applies with full force to genetic loci on different pairs of homologous chromosomes, genes housed on the same homologue often show independent or quasi-independent assortment as well. This is due to the phenomenon of crossing over.

During the early stages of meiosis, members of each pair of homologous chromosomes migrate to a side-by-side position in the cellular nucleus, and then routinely exchange portions of their DNA. The frequency of crossovers between pairs of linked loci varies from near zero for immediately adjacent genes to almost 100% for pairs of loci far apart on a long chromosome. Whenever a crossover event takes place, recombinant gametes are produced that typically have different multilocus allelic arrangements from their nonrecombinant (parental) counterparts (fig. 2.2). Predictable relationships exist between the frequencies of recombinant gametes and the physical distances between genes. Accordingly, geneticists can employ empirical estimates of recombination rates (as deduced

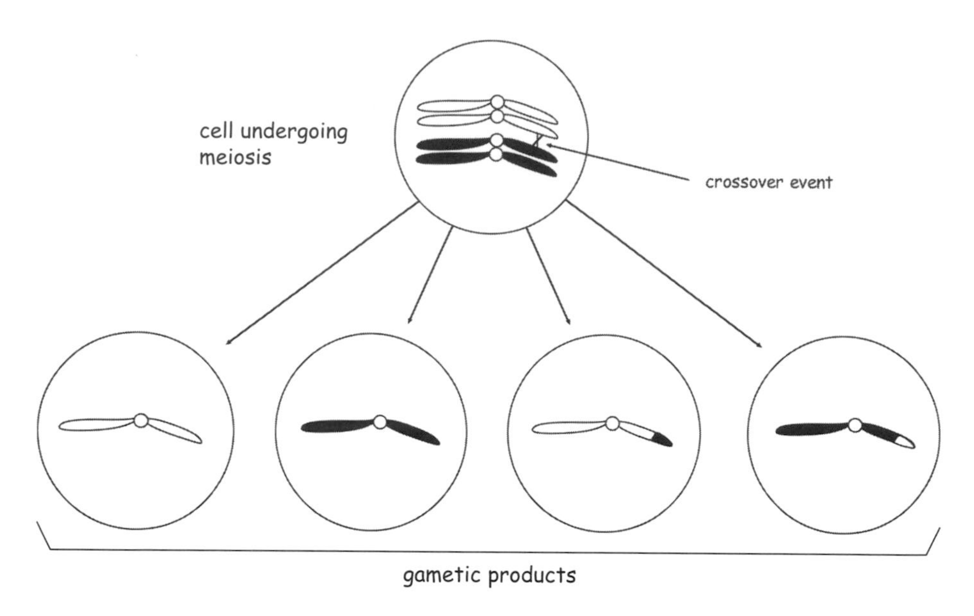

FIGURE 2.2 Alleles at linked nuclear loci are shuffled (recombined) by meiotic crossover events, which occur routinely during gametogenesis associated with sexual reproduction.

from suitable crossing studies) to map the relative positions of linked loci along a given chromosome.

Crossovers hence tend to shuffle the allelic contents of linked loci that otherwise would be inherited as a nonrecombined unit. Intrachromosomal crossover events merely add to what was already a vast scope for molecular recombination (interchromosomal) in the nuclear genome.

Recombination versus Mutation as Sources of Genetic Variation

The phenomena of sexual reproduction and genetic recombination are quite intimately wedded (as are their converses—asexual reproduction and clonality). In theory, recombination gives sexual populations far more adaptive flexibility than their asexual counterparts, all else being equal. However, the cellular processes of sex-based recombination would be genetically and evolutionarily inconsequential if natural populations carried little or no mutation-generated allelic variation to begin with. No amount of segregation, independent assortment, or syngamy can generate new genotypic combinations if allelic variation is unavailable for genetic shuffling. A hypothetical deck of 52 cards offers an analogy. If every card in the deck were a ten of hearts, even the most ardent shuffler could deal only identical hands. In sexual species, meiotic cells are deft in the mechanics of allelic shuffling, but their efforts would be to no avail if allelic variation were not already present in the deck of chromosomes from which they must deal.

Before the 1960s, evolutionary geneticists were uncertain whether natural populations harbored enough allelic variety to make genetic recombination a powerful force, relative to de novo mutation, in generating novel genotypic variations from one generation to the next. Then breakthroughs in molecular biology settled the issue unambiguously. Various molecular assays of proteins and DNA revealed that allelic variation in sexually reproducing species is normally extensive, such that recombination becomes fantastically powerful in generating new genotypic variety in each and every animal generation. Indeed, recombinant genetic variety in most sexual species is so pervasive that each individual is genotypically unique (box 2.1). This contrasts diametrically with the situation in asexual taxa, where many individuals can be clonally identical (especially in an

BOX 2.1 In Most Sexually Reproducing Species, Every Individual Is Genetically Unique

Every vertebrate animal has about 30,000 protein-coding genes (plus manyfold more noncoding DNA sequences). Molecular assays have amply demonstrated that most of these loci exist in multiple forms—different alleles—in a typical population. Thus the sexual mechanics of genetic recombination associated with meiosis and syngamy can generate vast arrays of different multilocus diploid genotypes.

This argument can be made quantitative. The rules of Mendelian heredity imply that recombination can theoretically generate $\geq 3^x$ different multilocus genotypes in a sexual population, where x is the number of loci with two or more alleles. (Three is the numeral raised to the xth power because at least three diploid genotypes [A1/A1, A1/A2, and A2/A2] are possible at each polymorphic locus.) If, for example, 30 genes are polymorphic in a sexual population, then the potential number of recombinant genotypes is at least 3^{30}, or 200 trillion. The numbers are vastly higher than this in most real species. In humans the genome is known to carry at least six million polymorphic sites, so the total possible number of genotypic combinations is at least $3^{6,000,000}$, or about 10 followed by 1.8 million zeros (Charlesworth, 2007). Such astronomically large numbers imply that no two people alive today (with the exception of monozygotic twins), or at any time in the past, or at any time in the foreseeable future, are likely to be completely identical genetically.

asexual taxon that is recent in origin and consequently has not yet had time to accumulate de novo mutational variation).

The Paradox of Sex

Sexual reproduction is an evolutionary enigma for several reasons: it dilutes (by 50%) a parent's genetic contribution to each offspring; it requires that each parent invest time and energy in finding a mate; it potentially exposes parents to sexually transmitted diseases; and it ensures that multilocus genotypes (all of which have passed the stringent tests of natural selection for proper functionality in parents) are thoroughly rearranged in progeny via the Mendelian genetic processes of segregation, independent assortment, and syngamy. All of these would seem to be inevitable costs of reproducing sexually, as opposed to clonally. Furthermore, these potential fitness costs are immediate rather than delayed, such that recombinational sex should in principle be highly exposed to negative natural selection in each and every generation.

By contrast, most of the potential fitness advantages of sexual reproduction seem at face value to be rather diffuse or postponed. For example, any sexual lineage, unlike an asexual one, has the potential to incorporate beneficial mutations that have arisen in separate individuals (Fisher, 1930; Muller, 1932), but because beneficial mutations are relatively rare, such allelic amalgamations are unlikely to pay reliable fitness dividends on a regular short-term basis. Additionally, although sexual reproduction certainly tends to break up multilocus combination of alleles that might interact poorly, whether this process can offer a routine, short-term fitness benefit to sex is problematic. And, if unopposed by natural selection, recombinational sex will also tend to quickly disassemble any favorable gene combinations that it may previously have forged.

Another potential advantage of sexual reproduction also appears on first inspection to be of long-term more than short-term benefit. In 1964, the famous biologist Hermann J. Muller suggested that harmful mutations accumulate by a ratchetlike mechanism in clonal lineages, thereby gradually lowering a population's mean genetic fitness (i.e., increasing mutational load) over evolutionary time. Most new mutations are deleterious rather than beneficial, so an asexual population can only ratchet downward in mean genetic fitness as it gradually accumulates such mutations. Calculations suggest that a strictly clonal population is unlikely to survive more than about 10^4–10^5 generations in the face of such incessant mutational pressure (Lynch and Gabriel, 1990; Gabriel et al., 1993). A sexual population, however, is less susceptible to Muller's ratchet because genetic recombination in the nuclear genome can help purge combinations of harmful alleles (Felsenstein, 1974; Kondrashov, 1988). Offspring with higher and lower mutational loads than the parents inevitably arise, and purifying selection maintains a manageable genetic load by eliminating unfit progeny. In principle, a sexual population of sufficiently large size can persist indefinitely with respect to circumventing the "mutational meltdown" (Lynch et al., 1993) otherwise expected for clonal reproducers under Muller's ratchet.

In strictly asexual genomes, all loci in effect are linked into one "supergene" transmitted as a unit across generations, and this can interfere in additional ways with the efficacy of natural selection. In theory, genetic linkage usually lowers the mean rate of incorporation of beneficial mutations into a population and increases the fixation rate of damaging mutations, in comparison to expectations for unlinked genes. Basically, this happens because any beneficial mutation in a gene tightly linked to other loci will necessarily drag along any deleterious alleles, at those other loci, with which it happened to be associated at the time of its origin. This population genetic phenomenon is variously referred to as "genetic draft" (Gillespie, 2000), the "hitchhiking process associated with selective sweeps" (Maynard Smith and Haigh, 1974), or the "Hill-Robertson effect" (Hill and Robertson, 1966). The reduced efficiency of natural selection in asexual populations, compared to sexual ones, is another reason to expect a long-term evolutionary advantage for sex.

Somewhat related to all of these notions is the idea that sexual reproduction accelerates the process of adaptation in ways that lower a population's risk of extinction. As phrased by Nick Colegrave, "sex releases the speed limit on evolution" (2002, p. 664). Most roads to extinction are long-term population-level processes, however, so this kind of argument appeals not as much to the differential fitnesses of individuals as to the differential extinctions of sexual versus asexual lineages. Although "group selectionist" arguments of this sort tend to be disfavored by evolutionary biologists (for good reasons), even ardent supporters of individual selection often acknowledge that sex may be a special case, and that species-level selection may indeed help to account for the disproportionate survival, and therefore representation in extant faunas, of sexual as opposed to clonal taxa (e.g., Fisher, 1930). Even if sexual reproduction does tend to promote long-term population survival, it should be remembered that genetic recombination on a generation-by-generation basis also inevitably produces many genetic disabilities and premature deaths (when some of the genotypes it randomly generates prove to be highly unfit).

Thus, overall, the obvious fitness costs of sexual reproduction appear to be immediate and high, whereas most of the benefits seem postponed or diffuse (although perhaps crucial to a population's long-term survival). This is the central enigma of sex: How has an evolutionary process that is guided by a myopic force—natural selection—promoted the prevalence of sexual reproductive modes whose most obvious fitness benefits appear to be deferred? Evolutionary biologists have long pondered this "queen of problems in evolutionary biology" (Bell, 1982, p. 19), and they have detailed their thoughts in many insightful papers and books (e.g., Ghiselin, 1974; Williams, 1975; Maynard Smith, 1978; Michod and Levin, 1988). Here I summarize standard conjectures about possible short-term advantages to sexual reproduction in higher animals. This is not to imply that longer-term or more diffuse advantages should be neglected, and indeed a pluralistic approach to understanding sex and recombination is probably necessary (Kondrashov, 1993; West et al., 1999).

In terms of habitat models, two types of scenario have been envisioned for how the immediate profits from sex might cover the fitness costs. Models of the "tangled bank" variety (Ghiselin, 1974) emphasize how spatial environmental heterogeneity may routinely favor the genetic diversity displayed among sexually produced progeny. The analogy to the tangled bank comes from the closing paragraph of *On the Origin of Species,* in which Charles Darwin mused about the amazing complexity of the organic world. Models of the "red queen" category (Van Valen, 1973) emphasize how short-term temporal variation in habitats might likewise favor the genetic diversity that characterizes sexual reproducers. The reference to the red queen comes from *Through the Looking Glass,* in which Lewis Carroll's Alice has to run as fast as she can merely to stay in place (the analogy being that populations must evolve rapidly simply to avoid extinction). The basic idea of both models is that environments are so heterogeneous, spatially or temporally, that populations in nearly every generation must have an adaptive flexibility that comes with the recombinational genetic variety borne of sexual reproduction. For example, vertebrate animals are routinely exposed to a vast array of disease-causing microbes, and so families and demes (local populations) presumably fare best when they continually display moving or variable genetic targets to the pathogens (Hamilton, 1980; Ladle, 1992).

Another hypothesis focuses on the debilitating effects of deleterious mutations. Some recent empirical assessments suggest that harmful de novo mutations arise so frequently in multicellular organisms as to pose consistent and immediate (as well as long-term) fitness challenges to natural populations (Haag-Liautard et al., 2007). If true, natural selection against deleterious mutations could be a routine proximate force, even on a generation-to-generation basis, favoring sex—that is, genetic recombination—as a means of countering Muller's ratchet.

A somewhat related but different set of scenarios focuses on mechanisms of DNA repair per se. Damages to genetic material (from environmental insults such as ultraviolet radiation, mutagenic chemicals, and oxygen radicals) are ubiquitous, with tens of thousands of new DNA lesions typically arising in each vertebrate cell every day. Especially for nuclear DNA, or nDNA, most of these damages are corrected immediately by a cell's restoration machinery, which consists of suites of enzymes engaged in DNA proofreading and repair. These intracellular repair systems are not infallible, for a few de novo mutations survive and replicate, but without their services multicellular life would be unsustainable.

A cell's mechanisms of DNA repair typically involve rebuilding a damaged piece of DNA using the intact information from a redundant, undamaged copy. One type of redundancy is the complementary strand in the double helix, which can serve as a template for DNA repair when the initial damage is confined to a single strand. Another source of genetic redundancy (in diploid organisms) is the second homologous copy of duplex DNA, which in a meiotic operation known as recombinational repair can provide an undefiled template for the rehabilitation of double-stranded nDNA damages. According to a theory promoted by Carol and Harris Bernstein (1991), all mechanisms for molecular recombination, including meiosis and syngamy in higher animals, represent evolutionary adaptations that originated in evolution; they are continually maintained by natural selection expressly for the functions they serve in providing the cellular machinery and proper genetic templates for the regular repair of nDNA damages.

The Bernsteins' argument generally proceeds as follows. In sexual species, a parent contributes only one cell to each offspring, so it is crucial that this gamete be as free as possible from genetic defects. The admixture of chromosomes from two separate individuals (a mother and a father) that takes place in each generation of sexual reproduction ensures a continuing source of undefiled nDNA template against which any damages to the homologous duplex can be repaired during meiotic recombination. Furthermore, outcrossing (as opposed to selfing) presumably is favored because it promotes the masking of deleterious mutations. Thus "DNA damage selects for recombination, and mutation in the presence of recombination selects for outcrossing" (Bernstein and Bernstein, 1991, p. 277).

Mitochondrial DNA is also subject to molecular lesions. Indeed, mtDNA molecules are housed in an intracellular environment where they are especially prone to damage from oxygen radicals generated by oxidative phosphorylation. Mammalian mtDNA, for instance, receives about 16-fold more oxidative damage on a per-nucleotide basis than does nDNA. Yet, ironically, animal mitochondria are thought to possess only limited DNA repair systems (this provides one explanation for why their DNA evolves so rapidly). Furthermore, mtDNA is packed with genes crucial to energy production in animal cells, and for this reason also it would seem to be desirable for organisms to have evolved refined mechanisms for mtDNA repair. Nonetheless, being asexually propagated through maternal lines, mtDNA apparently cannot avail itself of the meiotic recombinational repair that the Bernsteins argue is the primary mechanism of nDNA repair. Absent this capability, mtDNA must display some other means of avoiding Muller's ratchet. One hypothesis is that mtDNA molecules undergo nonmeiotic recombination, or gene conversion events, within the germ line in such a way that damage-free mtDNA templates correct faulty ones (Ogoh and Ohmiya, 2007; Tatarenkov and Avise, 2007). Another possibility (on which I will elaborate later in this chapter) is that mtDNA replication and molecular sorting during gametogenesis provide an alternative, nonrecombinational pathway for circumventing the evolutionary accumulation of genetic damages across animal generations.

Accordingly, the Bernsteins' DNA-repair scenario for the significance of meiotic recombination cannot be the full explanation for the proximate advantage of sexual reproduction, if for no other reason than that this gametogenetic process does not apply to clonally transmitted mtDNA. Furthermore, some empirical evidence has been interpreted to suggest that nDNA repair is not the direct purpose of meiosis, and questions also remain as to whether nDNA repair provides a credible explanation for outcrossing (see Kondrashov, 1993). But the DNA-repair hypothesis has been important because it suggests another plausible, and potentially testable, mechanism for how ongoing natural selection might consistently favor sexual reproduction—generation after generation—in vertebrates and other multicellular organisms.

Standing against all of the scenarios described above is an entirely different hypothesis: that sexual reproduction is, on average, often disadvantageous for higher animals in the short term, but that once evolved is now difficult to jettison. Perhaps, for example, sexual recombination arose in evolution and is maintained, at least in part, by natural selection acting on transposable elements or other genomic parasites that profit by being in host populations in which, by virtue of recombinational sex, the parasite's spread is facilitated (Hickey and Rose, 1988). Regardless of how recombinational mechanisms originated in evolution, there can be little doubt that many vertebrates are now mechanistically locked into the sexual reproduction mode, even if asexuality might otherwise benefit them over the short term (box 2.2).

To summarize this section, the three classical arguments for the evolutionary advantage of sexual reproduction—rapid adjustment to temporal or spatial habitat changes, facilitating the incorporation of beneficial mutations into a lineage, and expediting the removal of deleterious mutations—all seem to offer long-term but not necessarily short-term benefits to individuals and demes. Several plausible models (e.g., tangled bank, red queen, and DNA repair) have been advanced for how natural selection might routinely favor genetic recombination on a generation-by-generation basis, but these have proved difficult to evaluate critically. In the final analysis, no amount of armchair speculation can settle the issue of whether recombinational sex is of net short-term benefit or detriment, on average, to a multicellular animal in nature. This is why the few vertebrate species that do reproduce clonally are of special interest. Empirically, how well do these asexual lineages perform—ecologically and demographically—relative to their sexual cognates? And how long do clonal lineages typically persist in evolution, in absolute time and in comparison to sexual lineages? These subjects will be addressed in parts II and III of this book.

BOX 2.2 Many Vertebrate Species Are Evolutionarily Locked into Sexual Reproduction

One reason why sex is so prevalent among vertebrates and many other animals is that once sexual reproduction has evolved, it may be difficult or impossible to lose. For instance, essentially no mammals self-perpetuate asexually (see part II). This is due at least partly to the following reason, which involves an interesting phylogenetic legacy.

A peculiarity of mammals is that because of "genomic imprinting," an individual normally needs both paternal and maternal genes to develop successfully. During the imprinting phenomenon, which takes place in the course of gametogenesis, particular genes are chemically altered such that they can produce functional products only if they enter the zygote through a sperm cell, or, in other cases, only through an egg cell (Georgiades et al., 2001; Morison et al., 2005). When errors occur in this imprinting process, key genes fail to express properly, and death of the developing offspring typically ensues. For this mechanistic reason alone, if for no other, a female mammal would seem to have no option but to engage in sexual reproduction (Kono, 2006).

Other vertebrates do not exhibit genomic imprinting, so this phenomenon cannot provide a universal explanation for the evolutionary maintenance of sexual reproduction. Nevertheless, many other types of phylogenetic legacies—ranging from cellular-level processes to organismal-level behaviors—would probably make any reversion to asexuality mechanistically difficult, if not impossible, for most vertebrate species, even if clonality might otherwise be highly beneficial, at least in the short term.

Sex and Death

A long-held sentiment in both philosophy and science is that the phenomena of sexual reproduction and death are intimately related. The Nup people of Nigeria tell an evocative story that explains the inevitability of this relationship. In the beginning, God created a fixed number of stones, tortoises, and humans, each in two sexes (male and female). These individuals could not reproduce; instead, they merely reverted periodically to youth. Eventually the tortoises and humans decided they wanted children, and they asked for God's permission. God agreed, on the condition that their personal deaths must soon follow, lest the world become too crowded. So the adult tortoises and humans began to have children, and then soon passed away. Watching all of these proceedings, the stones decided not to have offspring, choosing instead to remain immortal as individuals. That is why, to this day, stones, unlike tortoises and humans, neither produce children nor die.

In the scientific literature, aging (or senescence) is defined as a persistent decline in the survival probability or reproductive output of an individual due to physiological deterioration. In other words, inherent biological fragility tends to increase with age and eventually culminates in death of the individual. There is little doubt that humans senesce: if the low death rate of 12-year-old children remained in effect indefinitely (implying that most deaths in older age classes were accidental rather than due to endogenous deterioration), then an average person would live to be approximately 1,200 years old and about one person per thousand would survive for at least 10,000 years. That fact that no such ancient individuals are alive today is compelling testimony that death probabilities increase drastically with a person's advancing age. Much the same can be said for other multicellular organisms, which appear to be inherently mortal rather than immortal.

The phenomenon of senescence has been modeled by evolutionary biologists and shown to be a mathematically logical consequence of the declining force of natural selection through successive age cohorts in a population. The basic argument (in simple verbal form) is as follows. Every gene in effect "wants" to propagate and spread through successive generations of its host species. The gene's odds are improved, all else being equal, if it promotes reproduction by young individuals more than by old ones, for at least two reasons: (1) in any generation or age cohort of animals, fewer and fewer individuals are alive in each successive time interval, if only because of accidental deaths; and (2) any gene that is consistently transmitted by the young in effect profits from the multigenerational "compounding of interest" that comes from having invested early, rather than late, in organismal reproduction. Thus, like the Nup god, genes tend to be indifferent to the personal well-being of seniors, in this case because the elderly in previous generations have had little impact, relative to the young, on a gene's representation in subsequent generations. In other words, natural selection disproportionately favors the perpetuation of genes that favor the young. Over evolutionary time, such genes inevitably tend to accumulate in populations.

This tendency has important evolutionary ramifications that are likely to play out in at least two ways. Under the "mutation accumulation" model, older age classes in a population become, in effect, genetic garbage bins where alleles with age-delayed deleterious somatic action concentrate in evolution simply because of weak selection pressure against their loss. A corollary of this model is that natural selection consistently favors any modifier gene that delays the ontogenetic expression of deleterious alleles, such that over evolutionary time the negative genetic effects also tend to gravitate into older age cohorts. Under the "antagonistic pleiotropy" model, alleles for senescence are favored by natural selection whenever their beneficial effects at early stages of life outweigh antagonistic harmful effects later on. For example, any allele that predisposes for the calcification of bones in adolescents might improve those individuals' mean genetic fitness by strengthening bodies, so such alleles would increase in frequency in the population even if, as a pleiotropic by-product, they also happened to promote atherosclerosis (the calcification of artery walls) later in life.

The hypotheses of mutation accumulation and antagonistic pleiotropy are not mutually exclusive, and both probably contribute to the aging phenomenon. But the basic concept they share—that genetic dispositions for senescence tend to evolve because younger individuals contribute disproportionately to the ancestry of future generations—seems logically irrefutable. This conclusion was first intuited more than a half-century ago when Peter Medawar (1957) noted that a relatively small advantage conferred early in the life of an individual may outweigh a catastrophic disadvantage withheld until later, and George Williams (1957) concluded that natural selection is biased in favor of youth over old age whenever a conflict of interest arises.

To recapitulate, the ultimate or most overarching evolutionary explanation for senescence and death in multicellular animals is the declining force of natural selection through successive age classes in a population. The penultimate evolutionary explanations then involve models such as mutation accumulation and antagonistic pleiotropy that describe how this age-related decline in selection intensity is likely to have been translated into genetic dispositions for senescence and mortality. Finally, at the proximate levels of physiological mechanism, the conventional kinds of medical explanations for aging (such as arthritis, cancer, heart disease, and a vast array of other age-related disabilities) come into play. Metabolic and physiological disorders may be the actual executioners of the elderly, but they are merely carrying out death sentences that have ultimately been handed down by long-standing judges and juries of natural selection in the powerful court of evolution.

Cellular Autonomy and Immortality

In vertebrates and other multicellular animals, individuals die but their genes may persist indefinitely, copied into descendant generations. Why do genes have a potential for immortality that is denied to multicellular individuals? The evolutionary explanations are multifaceted, but a key consideration (in my opinion) has frequently been overlooked: the notion of cell autonomy. The following brief account is taken from *The Genetic Gods: Evolution and Belief in Human Affairs* (Avise, 1998), which readers may wish to consult for additional background, details, and references.

Two distinct routes to genetic immortality—or at least to the persistence of replicating DNA across the eons—seem to have been available to life. The first pathway is clonal, and is approximated most closely by unicellular organisms with small genomes, such as some bacteria that reproduce mostly asexually. In such organisms, mitotic-type cellular proliferation can apparently outpace the accumulation of DNA damages and deleterious mutations, thereby circumventing Muller's ratchet and allowing a population to sustain indefinitely by "cellular replacement." (In truth, few bacterial taxa are strictly clonal, but their occasional engagements in recombination, via mating, transformation, or transduction, can be viewed as evolutionary insurance policies to the basic clonal tactic.) Formal models indicate that clonal populations of unicellular organisms can overcome Muller's ratchet only when genomes are very small. In such cases, some cells in each generation fortuitously escape de novo mutational damages; these become progenitors of the next generation. In a sense, then, each bacterial cell is potentially immortal because it can replicate itself to form a lineage. Indeed, the claim can be made that any extant bacterial cell has never experienced a personal "death" because "it" has been alive continuously across the eons.

The second pathway to immortality is blatantly sexual, and it is best exemplified by germ-cell lineages in multicellular organisms. In the much larger genomes that these lineages typically house, no cell is likely to escape mutational damage for long, given known mutation rates and the large numbers of nucleotide sites involved. So mechanisms of recombinational repair (as described above) probably operate in conjunction with gametic proliferation and intergametic selection (as described below) to counter DNA damages and deleterious mutations that would otherwise tend to accumulate across evolutionary time. All multicellular animals die, but their genes can live on indefinitely, perpetuated through cells of the germ line.

In both routes to genetic immortality, most of the potential gene-transmitting cells, either bacteria or gametes, die in each generation, but these sacrifices on the altar of natural selection do not typically compromise the genetic health and evolutionary continuance of compatriot cells that happen to survive. Thus, in unicellular creatures, each cell (i.e., organism) in a population is an independent functional agent of genetic transmission across the generations. In multicellular organisms, by contrast, only cells in the germ line have a somewhat analogous autonomy.

To emphasize why cellular autonomy is such a key concept in the contexts of organismal reproduction and personal death, consider a fanciful scenario: immortality for the multicellular animal. Even if some somatic cells and tissues could somehow keep pace with DNA damage via the asexual processes of cell replacement and turnover, this tactic could not confer immortality on the multicellular individual because the fates of any healthy somatic cells are tied inextricably to the remainder of the soma, which as a whole inevitably senesces and dies (as predicted by the evolutionary theories of aging discussed above). In other words, a severe malfunction of any crucial somatic tissue necessarily dooms *all* somatic cells in a multicellular animal, regardless of how healthy those remaining cells might be. For most multicellular organisms, the standard route of escape from this evolutionary predicament is, of course, the sexual strategy of producing autonomous germ-line cells. Gametes can be metaphorically interpreted as intergenerational lifeboats for genes that must flee somatic ships, all of which are guaranteed eventually to sink.

For the germ lines of multicellular animals, I propose that sexual *and* asexual tactics are simultaneously employed to generate gametes that can be relatively cleansed of genetic defects. Meiotic recombinational repair—a sexual tactic—helps to purge nDNA of molecular damages and deleterious mutations, and a molecular analogue of cellular replacement—an asexual tactic—helps to purge nonrecombining mitochondrial DNA of genetic liabilities. Thousands of mtDNA molecules populate most cells, so intense intracellular (i.e., intermolecular) selection pressures are expected to characterize the transmission of mtDNA from one cell generation to the next. Therefore, mtDNA molecules that survive and replicate to populate a mature oocyte must have been scrupulously screened by natural selection for replicative capacity, and (perhaps) for functional competency, assuming that incompetency often translates into cell deaths in the germ line. In effect, the molecular replacement process that applies to asexual mtDNA populations within eukaryotic cells may bear closer analogy to the cellular replacement process in clonal bacterial populations than it does to the recombinational aspects of molecular screening that apply to eukaryotic nDNA.

To the extent that either of these damage control tactics (recombinational repair and molecular replacement) fails before or during gametogenesis, the metabolic operations of germ-line cells may be compromised and gametic deaths may result. Tough screening by natural selection then continues at the zygote stage and during embryonic development, when the haploid genomes of two gametes that were thrown together during syngamy are first called upon to collaborate in diploid condition. Selection at these early life stages is intense, as evidenced by the fact that embryogenesis is a time of heavy mortality in most species.

In summary of this discussion about sex and life history, the Nup people were basically correct—the phenomena of sexual reproduction and death are intimately wedded. Many hundreds of millions of years ago, when multicellularity initially arose in evolution from ancestral unicellularity, the "die" was forever cast—individuals could no longer hope for personal immortality. With multicellularity inevitably came senescence and death, and also an evolutionary exigency for some genetic mechanism by which DNA could perpetuate itself despite each organism's impending demise. The successful solution that came to predominate is a sexual reproductive process entailing the generation and union of genetically recombined unicells (gametes) that can escape the parent individual's deteriorating multicellular somatic body. All of this happened, however, not by any direct edict from God, nor because of any ecological imperative to avoid overcrowding. Instead, mortality and recombinational sex in multicellular organisms ultimately exist because of natural selection's inevitable inattentiveness to the personal well-being of the elderly (as compared to the youthful).

This means, however, that mortality is inextricably associated with asexuality as well. The cells within each multicellular animal arise from a single precursor cell (the zygote) via a clonal process (mitotic cell division) that itself is essential for achieving high cellular kinship and the resulting intercellular collaboration necessary to produce a functional organism. From this perspective, clonality within the individual can be interpreted as a short-term opportunistic tactic for life. Multicellularity was a wonderful evolutionary invention, yielding individuals that can exploit countless ecological opportunities that are closed to unicellular organisms. However, multicellularity in animals is also a dead-end tactic when not accompanied by mechanisms that permit genes to survive the inevitable death of the individual. These additional mechanisms usually involve recombinational sex, but as we shall see in part II, they sometimes involve intergenerational clonality, too (albeit again through a single-cell stage).

SUMMARY OF PART I

1. Clonality and sexuality are polar opposites. Genes are faithfully copied during clonal replication, whereas they are genetically scrambled (recombined) during sexual reproductive modes. Therefore, any attempt to understand the ecological or evolutionary significance of sex is likely to be facilitated by analyzing clonal systems, and vice versa.

2. Clonality is often discussed in reference to whole organisms, but the phenomenon also applies to (and indeed is underlain by) genetic processes operating at the intraindividual levels of DNA molecules, chromosomes, cells, and genomes. All forms of clonal reproduction begin with the faithful replication of genetic material.

3. The clonal propagation of genetic material (nucleic acids) normally relies on the informational redundancy inherent in the double-stranded structure of duplex DNA. During mitosis, nuclear genomes are clonally copied during the proliferation of an individual's somatic cells, all of which trace back to a single cell (a fertilized egg or zygote in sexual species). A multicellular individual can hence be viewed as a tight-knit colony of clonemate cells.

4. Mitochondrial DNA—a genome in the cellular cytoplasm—is also clonally replicated during somatic cell proliferation, but the replication events are not closely

synchronized with cell divisions. Normally, mtDNA is also clonally transmitted across animal generations, through matrilines.

5. Sexual reproduction is the genetic antithesis of clonality because it involves the shuffling of DNA. In most multicellular organisms, this entails the production and subsequent union of gametic cells (egg and sperm) from two separate individuals. Meiosis is the cellular process by which nuclear genes are segregated and recombined during gametogenesis, and fertilization (syngamy) then consummates the broader recombinational operation by yielding offspring whose genotypes are different from one another and from those of both parents.

6. Sex has long been an evolutionary paradox because it would seem at face value to be highly detrimental, at least in the short term. Apparent costs of sex include the time and energy needed to find a mate; a potential exposure to sexually transmitted diseases; the breakup of gene combinations that have been selection-tested (in parents) and thus are likely to be beneficial to fitness; the cost (not incurred by clonal reproducers) of producing males; and a 50% dilution of one's nuclear genes in offspring (because the other 50% of an offspring's DNA has come from a genetically distinct sexual partner).

7. Sex is also a paradox because most of its advantages would seem at face value to be rather diffuse and postponed. These potential benefits include the incorporation, into a sexual lineage, of favorable mutations that arose in separate individuals; an amelioration or circumvention of Muller's ratchet (the expected evolutionary accumulation of detrimental mutations in clonal reproducers); the production of genetically diverse progeny, at least some of which might be suited to new environmental conditions; a potential for faster adaptive evolution (to keep pace, e.g., with ever-changing competitors, predators, prey, parasites, or pathogens); and in general, via any of the above-mentioned processes, a lower risk of population extinction over the longer haul.

8. Potential near-term benefits to sex have also been identified. Models of the "tangled bank" variety emphasize how spatial habitat heterogeneity might routinely favor the genetic diversity displayed among sexually produced progeny. Models of the "red queen" category emphasize how short-term temporal variation in environments might favor the genetic diversity that characterizes sexual reproducers. Another category of hypothesis focuses on the role of recombinational genetic processes in repairing the molecular damages that constantly plague DNA.

9. Ironically, the phenomena of senescence and death are inextricably wed, in evolution, to the phenomena of both sex *and* clonality. Clonal replication at the genomic and cellular levels permitted and fostered the evolution of multicellularity from unicellularity, but, in turn, multicellularity preordained senescence

and death of the individual (because of the inevitable declining force of natural selection in age-structured populations). And, once the phenomena of aging and death of the multicellular organism had arisen, sexual reproduction in effect became necessary as a means by which DNA could self-perpetuate indefinitely, by divorcing itself periodically from each sinking somatic ship. A seldom considered corollary is that regular passage through a unicellular (gametic) stage of the life cycle is a key aspect of the reproductive process for both sexual and clonal taxa.

10. Thus, within the multicellular individual, clonal replication of genomes (via mitosis) can be viewed as an opportunistic genetic tactic that can be of clear benefit in the short term, via the many ecological advantages that multicellularity may confer, but that is harmful in the slightly longer term in the sense of necessitating an individual's senescence and death. As we will see in part II, this theme of short-term evolutionary boon but longer-term bust reappears, albeit in different guises, at higher levels (clonal populations and "species") of the biological hierarchy.

PART II

Unisexual Clonality in Nature

Clonality is often associated with unisexuality—the presence of only females within a species. Indeed, the observation of a strongly female-biased sex ratio often provides the initial hint that a taxon under investigation might be a clonal reproducer. In 1932, Carl and Laura Hubbs described the first all-female vertebrate species known to science. It was a fish, which they named the Amazon molly in honor of a fabled human tribe of all-female warriors. The seemingly outlandish notion that these creatures reproduce clonally was later confirmed in the laboratory (Hubbs and Hubbs, 1946). Since then, dozens of all-female clonal or quasi-clonal taxa have been identified in various groups of fish, amphibians, and reptiles from five continents (Dawley and Bogart, 1989; Vrijenhoek et al., 1989). Collectively, these constitute about 0.1% of all extant vertebrate species.

It might appear that an absence of males in any unisexual taxon would sentence females to sexual abstinence and strict clonality, and indeed this is sometimes true. In other cases, however, females in a unisexual taxon mate with males of related bisexual (gonochoristic, or two-sex) species and then use the sperm to facilitate their own clonal or hemiclonal reproduction. Sperm-independent unisexuality (parthenogenesis) will be the topic of chapter 3, and sperm-dependent forms of unisexuality (gynogenesis, hybridogenesis, and kleptogenesis) will be described in chapter 4. All unisexual lineages of vertebrate animals appear to have arisen via hybridization between bisexual species. Thus taxa that are now asexual or quasi-asexual had long prior evolutionary histories of sexual reproduction, indicating that vertebrate clonality is a derived (not ancestral) as well as a polyphyletic (multiorigin) condition. In recent decades, much has been learned about the genetics, evolution, and ecology of these recurring transitions from sexuality to clonality.

Before moving to the chapters of part II, a brief introduction to nomenclatural practices for unisexual taxa is necessary. For sexually reproducing organisms, biological species are defined as groups of populations actually or potentially united by interbreeding—that is, by matings between genetically compatible males and females. In most clonal or quasi-clonal organisms, by contrast, there are no males and no genetically unifying mating events, so the traditional concepts and definitions of biological species do not apply. Primarily for this reason, any assignment of species names to unisexual taxa is problematic (Maslin, 1968; Cole, 1985; Frost and Wright, 1988), and the word "biotype" (rather than species) is sometimes preferred when referring to a particular collection of unisexual lineages. But asexual biotypes in nature often present themselves, empirically, as recognizable and quite discrete phenotypic entities, in part because of their recent hybrid ancestries; accordingly, taxonomists have often assigned Latin binomials to asexual biotypes, much as they do for sexual species. Although species-level taxonomies for unisexual biotypes are inherently somewhat arbitrary, they are nonetheless important because they inevitably influence our perceptions about clonal biodiversity and can even impact our conservation plans (Kraus, 1995).

CHAPTER THREE

Reproduction by the Chaste: Parthenogenesis

Standing inside her magnificent temple atop the Acropolis, the Parthenos once towered above her city as a silent protectress (fig. 3.1). This 40-foot-high statue of the goddess Athena, sculpted by Pheidias nearly 500 years before Christ, did not survive the Middle Ages; still, portions of the surrounding temple—the Parthenon—endure in Athens today as a powerful reminder of the glory of ancient Greece. "Parthenos" means virgin, but Athenians worshipped Athena as the mother of all men.

Virgins can indeed give birth, and not only in Greek mythology. In biology, the word "partheno" appended by "genesis" (meaning origin) describes the clonal process by which a female procreates without assistance from males or male gametes. She does so by producing special egg cells that require no sperm and no fertilization before initiating the development of offspring who, in most cases, are genetically identical to their virgin mother (fig. 3.2). This phenomenon exists in nature as a constitutive reproductive mode in several taxonomic groups of reptile. Each parthenogenetic biotype consists exclusively, or nearly so, of females who perpetuate their lines clonally, generation after generation.

The Cast of Players

Constitutive Parthenogens

The taxonomic order Squamata (lizards, snakes, and allies) provides all of the known vertebrate examples in which all-female clones are maintained in nature

FIGURE 3.1 Statue reconstruction of Athena Parthenos (in the National Archaeological Museum, Athens, Greece).

by parthenogenesis (Cole, 1975; Dawley and Bogart, 1989). Approximately 30 unisexual species have been formally named, representing about a dozen genera in more than half a dozen taxonomic families. No taxonomic family or genus is entirely parthenogenetic; each also includes bisexual species to which the parthenogens are evolutionarily allied and from which the unisexual clones arose via interspecific hybridization. Families containing parthenogenetic taxa are introduced in the following sections, and one parthenogenetic species representing each genus will be illustrated.

Lacertidae (Wall or Rock Lizards)

This family contains more than 25 genera and about 275 species native to Europe, Africa, and Asia. These small- to medium-sized terrestrial lizards have well-developed legs and a tail that is usually much longer than the body. They are diurnal (day-active), heliothermal (warmed by sunlight), mostly oviparous (egg-laying), and can be found in a wide variety of habitats, including forests, grasslands, scrublands, and rocky areas.

Most of these species are sexual, but a handful of asexual taxa in the genus *Darevskia* (formerly *Lacerta*) inhabit the Caucasus Mountains of eastern Europe, between the Black Sea and the Caspian Sea. Discovered in the late 1950s by Ilya Darevsky, these were the first unisexual reptiles known to science, and they have

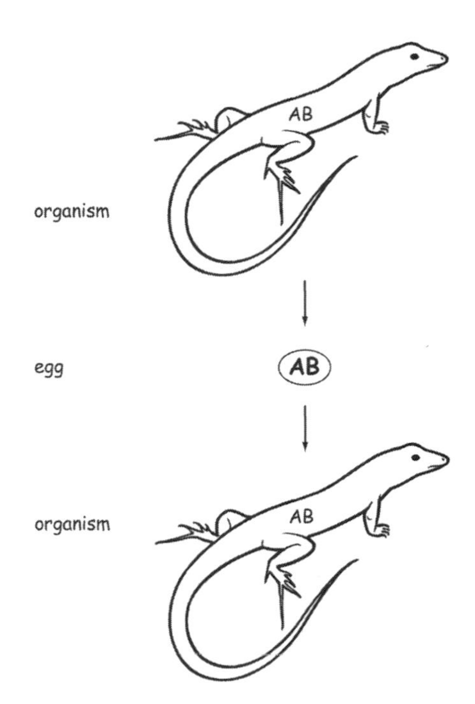

FIGURE 3.2 A simple summary of parthenogenesis. A and B refer to distinct nuclear genomes from two sexual species that hybridized to initiate the parthenogenetic lineage.

been the subjects of many genetic and ecological studies (Murphy et al., 2000). Clonal biotypes of *Darevskia* currently afforded formal recognition are *armeniaca, bendimahiensis, dahli, rostombekovi, sapphirina, unisexualis,* and *uzzelli*. More than a dozen other members of this genus are sexual (bisexual or gynodioecious, i.e., with two separate sexes).

Darevskia armeniaca

This parthenogenetic lizard (fig. 3.3) is known from scattered locations in northeastern Turkey, northern Armenia and Azerbaijan, and southern Turkey, within or near the ranges of the two sexual species (*D. valentini* and *D. mixta*) from which it arose via interspecific hybridization.

Gekkonidae (Geckos)

These familiar lizards are renowned for their flylike ability to walk across walls and ceilings, thanks to the presence in many species of expanded toes with padded lamellae that serve as miniature suction cups. They are also quite unique among lizards in their chirping vocalizations, rubbery-looking gray or tan bodies, and a transparent eye membrane that they lick to keep clean. Distributed worldwide in tropical and subtropical areas, the family contains more than 1,000 sexual species in approximately 100 genera.

FIGURE 3.3 *Darevskia armeniaca,* the Armenian rock lizard.

Most geckos are oviparous and show low fecundity. A female may produce just one clutch per year, typically consisting of only one or two hard-shelled eggs that she lays in a ground nest covered by organic material. Hatchlings sometimes remain near the nest for several years, until sexually mature. These mostly nocturnal animals are tolerant of a wide range of temperatures and can be quite abundant where suitable cover exists (including inside human habitations).

Five taxonomic species in three genera are known to be parthenogenetic, or at least to include parthenogenetic lineages: *Heteronotia binoei, Lepidodactylus lugubris, Hemidactylus garnotii, Hem. vietnamensis,* and *Nactus pelagicus.* Some of the unisexual biotypes have broad geographic distributions.

Heteronotia binoei

This taxon (fig. 3.4) will serve to introduce some biological and taxonomic complexities that often arise in discussions of unisexual biotypes and their sexual cognates. As traditionally recognized by taxonomists, *H. binoei* encompasses many bisexual as well as unisexual populations that collectively occupy much of Australia (Moritz, 1991, 1993; Strasburg and Kearney, 2005). Several geographic populations of the sexual reproducers differ in karyotype (gross chromosomal makeup), and two chromosomal races have hybridized to produce triploid clones that group into two distinct unisexual lineages informally known as 3N1 and 3N2. These asexual biotypes themselves have wide geographic distributions in western and central Australia.

FIGURE 3.4 *Heteronotia binoei,* Bynoe's gecko.

Lepidodactylus lugubris

This unisexual taxon (fig. 3.5) also has a huge geographic distribution, occurring on islands across much of the Indian and Pacific oceans. Collectively, *L. lugubris* consists of many parthenogenetic lineages, often sympatric, which can sometimes be visually differentiated by patterns of dark dorsal spots on the animals' bodies (Radtkey et al., 1995). Genetic data indicate that these clones arose through hybridization between the sexual species *L. moestus* (from Micronesia) and an undescribed sexual species whose range extends from French Polynesia to the Marshall Islands.

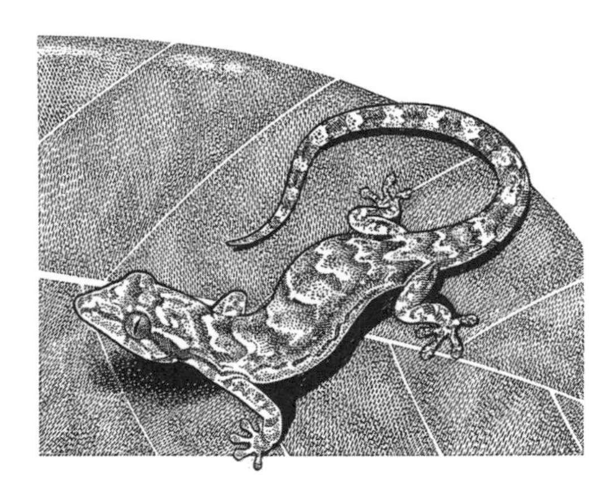

FIGURE 3.5 *Lepidodactylus lugubris,* the oceanian palm gecko.

FIGURE 3.6 *Hemidactylus garnotii,* the Indo-Pacific gecko.

Hemidactylus garnotii

This triploid unisexual species (fig. 3.6) is native to southeastern Asia and Indonesia, but has also been introduced elsewhere such as Florida, Puerto Rico, and Hawaii.

Nactus pelagicus

This taxon (fig. 3.7) is another example of a species complex that is comprised of both bisexual and unisexual populations (Donnellan and Moritz, 1995). Its range includes Queensland (Australia), New Guinea, the Solomon Islands, the Mariana Islands, Vanuatu, New Caledonia, Fiji, and Western Samoa. These geckos frequent forest edges, gardens, and coconut plantations.

FIGURE 3.7 *Nactus pelagicus,* the pelagic gecko.

Teiidae (Whiptail Lizards)

The Teiidae family contains more than 100 species, all confined to the Americas. These species are diurnal, carnivorous, oviparous, and mostly terrestrial, although a few species are semiarboreal or semiaquatic. They have strong limbs and are fast runners. They also have scaly eyelids, conspicuous ear openings, pointed heads, elongate bodies and tails, and bodies usually patterned by stripes, checks, crossbars, or spots. Most species are bisexual, but about 15 unisexual biotypes—some diploid and some triploid—are known within the family. The phylogeny and systematics of Teiidae have been addressed extensively by Charles Cole, Herbert Dessauer, and their colleagues at the American Museum of Natural History, and in what follows I will employ the taxonomy and classification they recently recommended (in Reeder et al., 2002).

The genus *Aspidoscelis* (formerly *Cnemidophorus,* in part) includes about ten recognized taxa that are either parthenogenetic or include parthenogenetic lineages (Wright and Vitt, 1993), an example of which is detailed below. These animals inhabit desert regions mostly in the southwestern United States, and many have been studied extensively in terms of ecology, genetics, and behavior by several top researchers (box 3.1). Interestingly, some of the unisexual females participate in pseudocopulations with one another (Crews and Fitzgerald, 1980). These mating behaviors are probably an evolutionary holdover from sexual ancestry, but data indicate that they also serve as copulatory stimuli that enhance the unisexuals' reproductive output (Crews et al., 1986).

The family Teiidae includes three additional genera—*Kentropyx, Teius,* and *Cnemidophorus* (sensu stricto)—with known parthenogenetic representatives. An example is the diploid keeled tegu *K. borckiana* (fig. 3.8) from northern South America, which arose from a cross between the gonochoristic species *K. striata* and *K. calcarata* (Reeder et al., 2002). The unisexuals *Teius suquiensis* and those in the *Cnemidophorus lemniscatus* complex are also South American (Sites et al., 1990; Vyas et al., 1990; Avila and Martori, 1991).

Aspidoscelis uniparens

This species (fig. 3.9) has a rather restricted range confined to portions of Arizona, New Mexico, extreme western Texas, and northern Mexico. All of the other unisexual species within the genus have similarly narrow distributions in southwestern regions of North America.

Gymnophthalmidae (Spectacled Lizards)

Native to Central and South America, this family contains nearly 200 bisexual species in about 40 genera. Various species can be found from arid deserts to tropical rainforests and from low elevations to high-altitude paramos in the Andes.

BOX 3.1 Wesley Brown and Craig Moritz

In 1968, Wesley Brown entered graduate school at the California Institute of Technology in Pasadena, where he was introduced to mitochondrial DNA in the laboratories of Giuseppe Attardi and Jerome Vinograd. These researchers studied mtDNA transcription and physical chemistry, respectively, but Brown's interests were at least as much in the area of natural history. In 1971, he went to an exhibition of Max Escher paintings at the Los Angeles County Museum, where he happened to meet John Wright, curator of the herpetology department. Wright was one of the most knowledgeable people in the world on *Cnemidophorus* lizards (he later coedited a book on the biology of these creatures), and Brown's chance meeting with him that day was to lead to their collaborative studies on the evolutionary origins of parthenogenetic taxa from a genealogical perspective.

For the next few years, Brown gathered mtDNA data for various projects, and in 1975 he published an article with Wright entitled "Mitochondrial DNA and the Origin of Parthenogenesis in Whiptail Lizards (Genus *Cnemidophorus*)." This paper, and a 1979 follow-up published in *Science,* were classics in at least two regards: they were among the first applications of mtDNA data to any natural population, and they provided the first molecular documentation of genealogy in any unisexual species. In those papers, Brown and Wright deduced from mtDNA and other genetic evidence that the unisexual taxa *Cn. neomexicana* and *Cn. tesselata* originated relatively recently in evolution by hybridization between female *Cn. marmorata* and males of *Cn. inornata* and *Cn. septemvittata,* respectively. This was a remarkable achievement, totally unanticipated. In 1978, just one year before the *Science* paper appeared, a leading evolutionary biologist (M. J. D. White) had lamented in print that for parthenogenetic taxa we are never likely to know which species was the female parent (p. 61), because there seemed no way to retrieve such information before maternally transmitted mtDNA was discovered.

Meanwhile, in his native Australia, graduate student Craig Moritz was beginning to publish on the ecology and cytology of parthenogenetic lizards (*Heteronotia*) on that continent. In the early 1980s, his interest in parthenogenesis drew him to Wes Brown's lab, then at the University of Michigan. There, as a postdoctoral student, Moritz learned and applied mtDNA methods to evolutionary questions about additional *Cnemidophorus* unisexuals and their bisexual relatives. The result was a classic series of papers (many coauthored with Lou Densmore) on the origin and relative age of these North American species. After completing this work, Moritz returned to Australia to continue, in Melbourne, genetic analyses of other unisexual taxa; later, he moved to the University of California at Berkeley. Over the years, he and his colleagues have produced more than 20 scientific publications on the evolutionary genetics of vertebrate parthenogenesis.

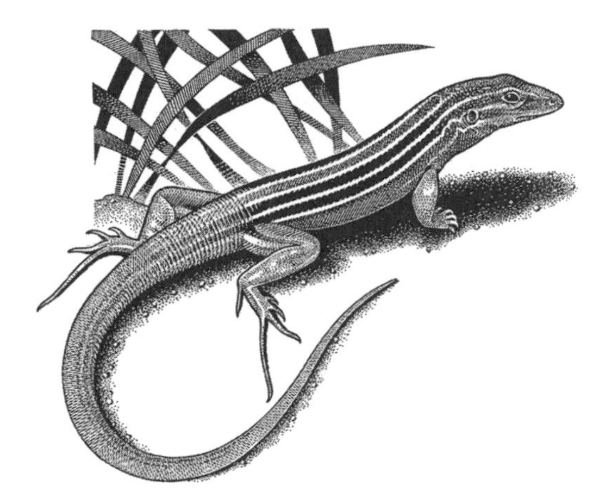

FIGURE 3.8 *Kentropyx borckiana,* the keeled tegu.

FIGURE 3.9 *Aspidoscelis uniparens,* the desert grassland whiptail.

Most of these animals have legs, but limb reductions (accompanied by body elongations) have occurred several times, and a few species are limbless burrowers in soils of tropical forests. Gymnophthalmidae and Teiidae are closely related and may be sister taxa, tracing to the same ancestral lineage. Indeed, gymnophthalmid species were formerly placed in the family Teiidae.

Within the Gymnophthalmidae, unisexual parthenogenesis apparently characterizes some populations of *Gymnophthalmus underwoodi* (fig. 3.10) in northern South America and Brazil (Hardy et al., 1989; Cole et al., 1990, 1993; Yonenaga-Yassuda et al., 1995; Kizirian and Cole, 1999). Parthenogenesis is also known in various populations of *Leposoma percarinatum* (fig. 3.11) in South America (Uzzell and Barry, 1971; Pellegrino et al., 2003).

Xantusiidae (Night Lizards)

Many of these viviparous (live-bearing) animals are actually diurnal, although they tend to be highly secretive in their North and Central American haunts. They have large platelike scales on their heads, soft and fragile skin that is loose-fitting, an easily broken tail, conspicuous ear openings, vertical pupils, and transparent eyelids.

The family contains three genera and about 20 species, one of which (*Lepidophyma flavimaculatum;* fig. 3.12) reportedly includes parthenogenetic all-female populations in a lower Central American portion of the species range (Sinclair et al., 2006).

Scincidae (Skinks)

Skinks are rather elongate lizards with relatively long snouts, flattened skulls, and smooth and shiny scales. This cosmopolitan group, with approximately 1,200 extant species in 130 genera, is a recent addition to the list of lizard families displaying parthenogenesis (Adams et al., 2003). Based on molecular markers combined with evidence on karyotypes, sex ratios, and inheritance data, the widespread

FIGURE 3.10 *Gymnophthalmus underwoodi.*

FIGURE 3.11 *Leposoma percarinatum.*

FIGURE 3.12 *Lepidophyma flavimaculatum,* the yellow-spotted night lizard.

Australian lizard *Menetia greyii* (fig. 3.13) is actually an assemblage of sexual populations *and* closely related parthenogenetic lineages (Adams et al., 2003).

Typhlopidae (Blind Snakes)

These burrowing snakes, with vestigial scale-covered eyes and a rounded head, look almost like earthworms. They comprise a family of six genera and about

FIGURE 3.13 *Menetia greyii,* the common dwarf skink.

240 species found mostly in tropical regions of Asia, Africa, and the Americas. One southeastern Asian species (*Ramphotyphlops braminus;* fig. 3.14) reportedly consists of parthenogenetic females, and is the only documented instance of constitutive parthenogenesis in a nonlizard vertebrate.

Other Taxa

A few additional reports of suspected constitutive parthenogenesis exist for vertebrate animals, typically based on the observation of a female-biased sex ratio in a natural population, or on the presence of many triploid individuals that characteristically are common in parthenogens but rare in sexual taxa. (All triploid lineages of reptile in nature have been thought to be parthenogenetic [Suomalainen et al., 1987]), but caution is indicated because a few triploid forms of frog, in the genus *Bufo,* were recently discovered that reproduce bisexually [Stöck et al., 2002, 2005].) Among reptiles, additional examples of suspected parthenogens are triploid lizards (*Leiolepis triploida*) of the family Agamidae and a subspecies of chameleon (*Brookesia spectrum affinis*) in Chamaeleonidae (Hall, 1970). These taxa have not been studied sufficiently to merit further consideration here.

Although not the focus of this book, many invertebrate taxa also display parthenogenesis or other related forms of clonality or hemiclonality. A brief introduction to these creatures is presented in box 3.2.

FIGURE 3.14 *Ramphotyphlops braminus,* the brahminy blind snake.

Sporadic Natural Parthenogens

All of the unisexual biotypes described above perpetuate clonal lineages by obligatory or constitutive parthenogenesis. A different form of parthenogenesis (tychoparthenogenesis, or facultative parthenogenesis; Simon et al., 2003) occurs sporadically in a few sexual vertebrates. Zookeepers have long suspected that some bisexual species are capable of tychoparthenogenesis, based on an observation that females maintained in long-term isolation occasionally produce viable offspring. Molecular genetic assays have confirmed that some such progeny do carry their mother's DNA only, and thus indeed had a single parent.

In some cases, these offspring have proved to be clonally identical to their mother and to one another. Perhaps the best-documented example involves a Burmese snake, *Python molurus bivittatus*. One virgin female in an Amsterdam zoo generated embryo-containing eggs for five consecutive years. DNA fingerprinting revealed that these all-female progeny were genetically identical to their mother, thus confirming clonal parthenogenesis and eliminating a competing hypothesis that the mother had utilized sperm from a clandestine mating (Groot et al., 2003).

In two studies involving other bisexual snake species, occasional progeny from supposed virgins also proved upon genetic examination to be of single-parent genesis, but the animals in these cases were not genetically identical (Dubach

BOX 3.2 Invertebrate Parthenogenesis

Parthenogenesis-type modes of reproduction are represented in many invertebrate phyla, including Cnidaria, Platyhelminthes, Nematoda, Rotifera, Echinodermata, Mollusca, Annelida, and Arthropoda (Suomalainen et al., 1987). The phenomenon is rare in some groups, such as Echinodermata—where perhaps only one parthenogenetic biotype, *Ophidiaster granifer,* has been reported—but in various other groups the phenomenon is relatively widespread. In arthropods, for example, parthenogenesis occurs across a diversity of crustaceans, insects, arachnids, and other taxa.

The collective diversity of parthenogenetic modes is greater among invertebrates than among vertebrate taxa (Beukeboom and Vrijenhoek, 1998). In terms of genetic mechanisms, various invertebrate parthenogens reproduce either clonally or nonclonally depending on the cytogenetic mechanism (e.g., meiotic versus nonmeiotic) of egg production in the taxon under consideration. With respect to life history, some invertebrate taxa are *cyclical* rather than obligate parthenogens, meaning that successive generations periodically alternate between sexual and parthenogenetic modes, usually as a function of ecological circumstance. Some species of *Daphnia* (small water flea crustaceans), for instance, tend to reproduce parthenogenetically when in stable bodies of water but via sexual reproduction when in ephemeral habitats (Lynch, 1983, 1984a). This kind of cyclical parthenogenesis is unknown in vertebrates. And in terms of evolutionary origin, all unisexual vertebrates seem to have arisen by interspecific hybridization, whereas some invertebrate parthenogens possibly have not (Beukeboom and Vrijenhoek, 1998).

Another seeming contrast is that some invertebrate parthenogenetic clades—but no known vertebrate ones—are suspected to be evolutionarily ancient (that is, at least tens of millions of years old). Three plausible examples are darwinulid ostracods (Martens et al., 2003), bdelloid rotifers (Mark Welch and Meselson, 2000; Mark Welch et al., 2004), and orbatid mites (Heethoff et al., 2007). If these groups truly have persisted across the eons effectively without sex (this topic is controversial; see Butlin, 2000; Avise, 2004a, pp. 179–180; Domes et al., 2007), they would be evolutionary scandals (Maynard Smith, 1986; Judson and Normark, 1996) that contravene the conventional wisdom that multicellular organisms require genetic recombination for long-term evolutionary persistence.

et al., 1997; Schuett et al., 1997, 1998). Thus a nonclonal (i.e., meiotic) form of sporadic parthenogenesis had transpired. (Actually, because unions of haploid cells were involved, this reproductive mode could also be considered a form of hermaphroditic self-fertilization, a topic otherwise deferred to chapter 6.) These offspring were also highly homozygous, implying that the cytological mechanism likely involved a fusion of identical haploid nuclei within a female such that the resulting progeny in effect were inbred siblings rather than clonemates. One likely hypothesis is that this outcome arose by a form of automixis, or automictic parthenogenesis, wherein mitotic-type divisions follow a meiotic reduction division and yield multiple haploid cells, some of which are genetically identical. If two such clonemate cells successfully unite, the net result would be an offspring that is "instantly inbred" and highly homozygous, yet different in genetic composition from its parents and its siblings.

To nearly everyone's surprise, a similar example of facultative meiotic parthenogenesis was recently uncovered in captive populations of monitor lizards (Lenk et al., 2005) including the Komodo dragon (Watts et al., 2006), one of the world's most impressive lizards (fig. 3.15). Interestingly, all of the offspring were male—not female—and highly homozygous. This sex bias likely reflects the fact that in Komodo dragons, the male is the homogametic sex (i.e., ZZ) whereas females are heterogametic (i.e., ZW). Thus only males would arise under the mechanism of automixis described above.

Chapman and colleagues (2007) uncovered another unexpected example of facultative automictic parthenogenesis, this by a captive specimen of *Sphyrna*

FIGURE 3.15 *Varanus komodoensis,* the Komodo dragon.

FIGURE 3.16 *Sphyrna tiburo,* the bonnethead shark.

tiburo, the bonnethead shark (fig. 3.16). This was the first report of parthenogenesis in the class Chondrichthyes, which includes cartilaginous fish such as sharks, rays, and chimeras.

Tychoparthenogenesis has long been known in a few other vertebrate species as well, for example turkeys and other domesticated game birds (Olsen, 1974). How often facultative parthenogenesis occurs in nature remains unknown, but instances of the phenomenon seem plausible, especially in taxonomic groups (notably reptiles) in which the physiological capacity for virgin births is known to be widespread. Documenting the frequency of facultative parthenogenesis in the wild will remain challenging, however. It typically necessitates genotyping many progeny for which the dam is known, identifying progeny that appear to carry only alleles that could have been inherited from their mother, and eliminating the possibility that the dam had mated with a male who shared, perhaps by virtue of close kinship, her alleles at the particular loci surveyed. Furthermore, mere observation of continued reproduction by a dam who is isolated from males is insufficient to verify parthenogenesis, because females in many reptilian species can store and utilize viable sperm for long periods of time—sometimes up to several years—following a successful mating (Birkhead and Møller, 1993; Pearse et al., 2001).

In any event, facultative parthenogenesis in vertebrates seems to be a relatively unusual and ephemeral phenomenon, typically confined to occasional instances of single-generation uniparentage and sometimes, but not always, entailing clonality. We will henceforth confine attention in this chapter to parthenogenetic lineages that clonally self-perpetuate in nature.

Additional Parthenogens?

Although new reports of parthenogenetic lineages continue to appear (e.g., Rocha et al., 1997; Adams et al., 2003), the pace of discovery of vertebrate parthenogens (and other unisexual biotypes to be discussed in chapter 4) has slowed in recent decades compared to the heyday era of the late 1950s to the early 1980s. Perhaps most instances of vertebrate unisexuality have already been documented. However, two recent biases against the detection of constitutive clonality might also be involved. First, the traditional kinds of data initially suggestive of unisexuality—female-biased sex ratios and natural history details—seem to be gathered less often now because museum workers and systematists generally tend to collect fewer vertebrate specimens. This restraint is due to ethical concerns about declining biodiversity, as well as to stricter laws and protective regulations for vertebrate animals. Second, karyotypic and protein-electrophoretic assays that were previously popular in population genetic analyses have been frequently supplanted in recent years by surveys of mtDNA, a maternally inherited molecule that cannot by itself be used to distinguish clonal from sexual lineages (see beyond). Thus many additional examples of constitutive parthenogenesis and other forms of clonality and quasi-clonality in vertebrates undoubtedly await scientific discovery.

Cellular and Genetic Mechanisms

In terms of genetic outcome, constitutive parthenogenesis can be likened to the operation of mitosis extended across animal generations. However, the chromosomal mechanics, where known, are also quite similar to meiosis (fig. 3.17). In the parthenogenetic lizard *Aspidoscelis.* (formerly *Cnemidophorus) uniparens,* the process begins with an endomitotic event in which all of a cell's chromosomes replicate without cell division, after which the identical chromosomes synapse (align), engage in crossing over, and segregate into ova (Cuellar, 1971). These cellular maneuvers are reminiscent of standard sexual meiosis except that they do not produce recombinant genetic variety, because the synapses, crossovers, and resultant segregations involve duplicated chromosome pairs rather than nonsister homologues. So this parthenogenetic operation has elements of both meiosis and mitosis, and is appropriately termed premeiotic endomitosis. The resulting unreduced ova are genetically identical to the somatic cells of the mother, yet are capable of dividing mitotically to produce a new daughter individual.

For vertebrate taxa that are constitutively parthenogenetic and have been examined in cytological detail (Uzzell, 1970), premeiotic endomitosis seems to be the usual cytogenetic mechanism by which all-female lineages propagate. By contrast, parthenogenetic plants and invertebrate animals collectively display

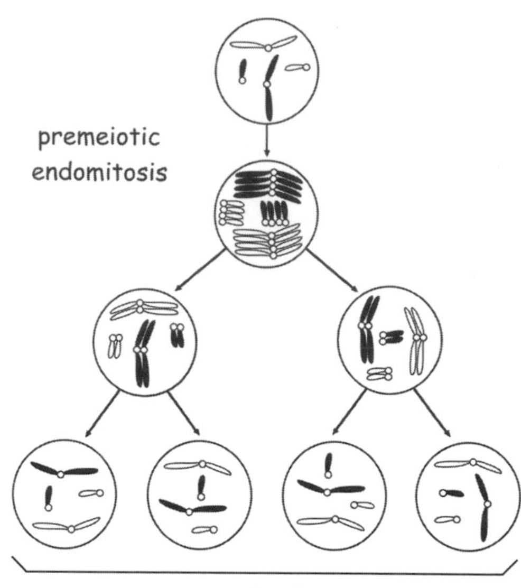

FIGURE 3.17 The parthenogenetic process of premeiotic endomitosis.

a broader array of cellular mechanics (Suomalainen et al., 1987; Bullini, 1994), also including meiosis followed by fusion of genetically different haploid nuclei from the same parent (another form of automixis or meiotic parthenogenesis), or the full abandon of meiosis in favor of strict mitosis (ameiotic or apomictic parthenogenesis).

Evolution and Phylogeny

Hybrid Origins

In all known cases of constitutive parthenogenesis in vertebrates, each clonal lineage originated via interspecific hybridization (fig. 3.18) and is therefore heterozygous at most nuclear loci that distinguish its ancestral sexual species (e.g., Cole et al., 1983, 1988; Good and Wright, 1984; Dessauer and Cole, 1986). The fact that each cell in a parthenogenetic animal houses at least two distinct nuclear genomes probably explains why the chromosome sets fail to synapse during oogenesis and, in general, why a dysfunction exists in the normal process of meiosis (Neaves, 1971; Cole, 1975; Darevsky, 1992). In other words, a cause-and-effect relationship is implicated between hybridization and the spontaneous origin of parthenogenesis in particular lineages of squamate reptiles. Some

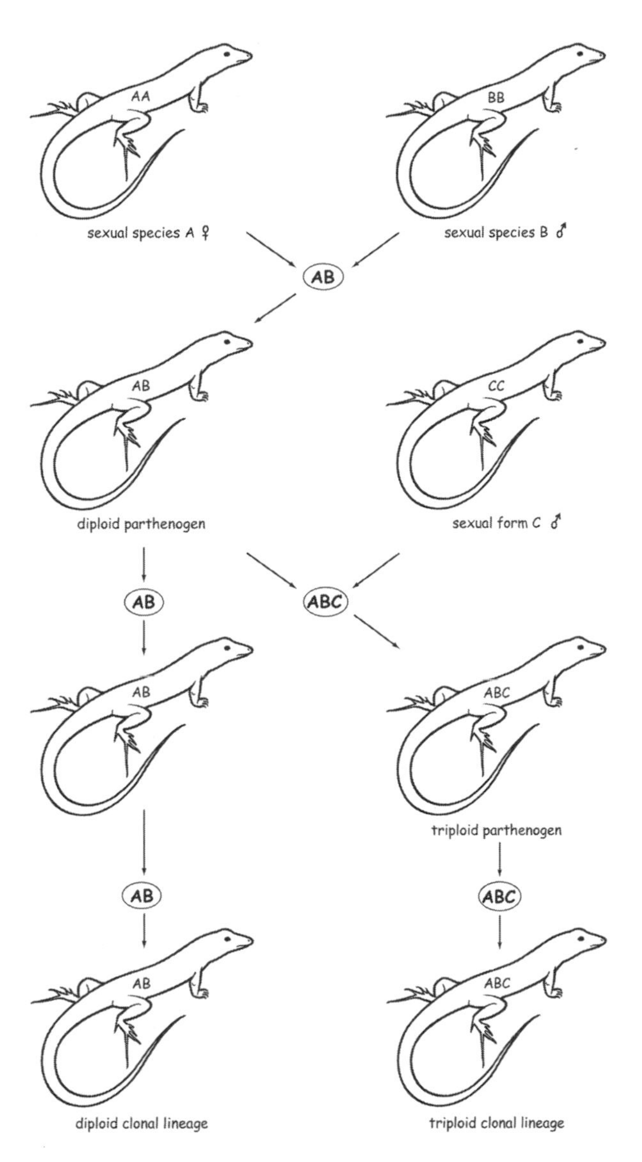

FIGURE 3.18 Types of hybridization events involved in the evolutionary genesis of diploid and triploid parthenogenetic lineages (after Dawley, 1989). These scenarios apply, for example, to parthenogenetic lineages of the diploid *Aspidoscelis tesselata* and triploid *A. neotesselata* complexes, where A, B, and C refer to haploid nuclear genomes that trace genealogically to the ancestral sexual taxa *A. tigris marmorata, A. gularis septemvittata,* and *A. sexlineata viridis,* respectively. In general, C can also refer to a third contributing individual that is conspecific to A or B.

parthenogens are diploid and clonally perpetuate nuclear genomes from the two sexual ancestral species from which they arose. Other parthenogens are triploid and clonally perpetuate three nuclear genomes via unreduced triploid eggs. The three genomes of a triploid parthenogen often trace to two ancestral species (in a 2 + 1 configuration); sometimes, though, they derive from three distinct sexual species (in a 1 + 1 + 1 configuration) that engaged in two successive hybridization events, probably as diagrammed in figure 3.18.

Parent Sexual Species and Direction of Cross

Scientists routinely use molecular markers from nuclear and mitochondrial DNA, typically in conjunction with morphological and distributional evidence, to deduce the sexual parent species and the direction of cross that gave evolutionary rise to particular parthenogenetic biotypes (box 3.3). From a compilation of such cytonuclear appraisals (table 3.1), several points have emerged. First, the hybridization event(s) that initiated an extant parthenogenetic taxon typically were unidirectional with respect to sex. (A standard caveat, however, is that more extensive population sampling might yet reveal extant parthenogens from the reciprocal cross.) Reciprocal hybridization events have been documented in only a few instances: for example, parthenogenetic *D. uzzelli* arguably arose via crosses in both directions with respect to sex (see footnote to table 3.1); parthenogenetic lineages in the *H. binoei* complex collectively register reciprocal crosses between distinct sexual races (CA6 and SM6) of *H. binoei;* and a diploid hybrid that was an intermediate ancestor of several triploid lineages of *Aspidoscelis* lizards (*exsanguis, flagellicauda*-like forms, and *opatae*-like forms) was deduced to have arisen via reciprocal hybridization events between *A. inornata* and probably *A. burti* (Reeder et al., 2002).

Second, the genetic appraisals have demonstrated that some parthenogenetic lineages that have been assigned different Latin names (such as *A. sonorae* and *A. flagellicauda*) share the same parent species and direction of cross despite clearly having originated in separate hybridization events. Third, some parthenogenetic biotypes currently included under a single Latin binomial, such as *D. uzzelli* or *H. binoei,* include multiple lineages that arose by different hybridization routes (table 3.1).

The parent species that hybridized to produce each unisexual biotype, although invariably congeneric (in the same taxonomic genus), typically have proved to be somewhat distant genetic cousins rather than closest evolutionary relatives (Moritz et al., 1992a, 1992b; Murphy et al., 2000). This pattern is illustrated in figure 3.19 and figure 3.20 for the two lizard genera—*Aspidoscelis* and *Darevskia*—that have been studied most thoroughly in these regards. Such phylogenetic outcomes cannot be attributed to a lack of hybridization between sister species or other near phylogenetic kin; hybrid animals from such close crosses

BOX 3.3 Cytonuclear Dissections of Unisexual Origins

Unisexual lineages in vertebrates seem invariably to have arisen by hybridization events between sexually reproducing species, and molecular genetic appraisals can help to identify the parent species and the direction of hybrid cross in each instance. The straightforward logic of cytonuclear analysis is illustrated in the table below, the body of which lists alleles at various loci in five hypothetical sexual species that were plausible candidate parents for parthenogenetic lineage "X." First, data from several diagnostic unlinked nuclear loci (*nl*) are used to pinpoint the sexual species (*sp*2 and *sp*5 in this case) that were ancestral to the parthenogenetic biotype. Then mitochondrial (*mt*) markers are used to identify the maternal parent species (*sp*5), thereby also making evident the original paternal parent (*sp*2).

	nl1	*nl2*	*nl3*	*nl4*	*nl5*	*nl6*	*mt*
Candidate sexual parents							
*sp*1	A	C	D	B	A	D	A
*sp*2	A,B	F	D,E	C	A	B	B
*sp*3	A,C	D,F	A,C	D	B	A,D	C
*sp*4	B,C	A	B	A,B	D	B,D	D
*sp*5	D,E	C	C	A,C	E	A,D	E
Parthenogenetic lineage "X"	A/E	C/F	C/D	C	A/E	B/D	E

are sometimes observed in nature, but they normally prove to be sexual rather than parthenogenetic (e.g., Walker 1981a, 1981b; Dessauer et al., 2000). Examples involving *Darevskia* lizards are presented in table 3.2.

Two hypotheses have been advanced to account for why hybridization events between phylogenetically distant (rather than closest) pairs of congeners give rise to parthenogenetic lineages. Under the balance hypothesis, parthenogenesis can arise only when the genomes of parental species are divergent enough to disrupt meiosis in hybrids, yet not so divergent as to seriously compromise hybrid viability or fertility (Moritz et al., 1989a; Vrijenhoek, 1989a). Under the phylogenetic constraint hypothesis, genetic peculiarities happen to predispose particular parental species to produce parthenogenetic lineages when they hybridize (Darevsky et al., 1985). These two hypotheses are not mutually exclusive.

Craig Moritz and colleagues interpreted phylogenetic findings for *Aspidoscelis* (*Cnemidophorus*) as consistent with the balance hypothesis, given that several

TABLE 3.1 Examples of parthenogenetic biotypes for which the parental species and direction(s) of the original hybrid cross(es) have been identified using combinations of nuclear and mtDNA markers[a]

		Sexual parental species (subspecies)		
Parthenogenetic biotype	*Ploidy level*	*Males*	*Females*	*References*
Aspidoscelis (formerly *Cnemidophorus*) lizards				
cozumela	2N	*deppii*	*angusticeps*	Fritts (1969); Moritz et al. (1992b)
rodecki	2N	*deppii*	*angusticeps*	Moritz et al. (1992b)
laredoensis	2N	*sexlineata*	*gularis*	Wright et al. (1983); Parker et al. (1989)
neomexicana	2N	*inornata*	*tigris* (*marmorata*)	Brown and Wright (1979); Densmore et al. (1989b); Dessauer et al. (1996)
tesselata	2N	*gularis* (*septemvittata*)	*tigris* (*marmorata*)	Brown and Wright (1979); Densmore et al. (1989b); Dessauer et al. (1996); Walker et al. (1997)
neotesselata	3N	*gularis* (*septemvittata*); *sexlineata* (*viridis*)	*tigris* (*marmorata*)	Brown and Wright (1979); Densmore et al. (1989b); Dessauer et al. (1996); Walker et al. (1997)
flagellicauda	3N	*burti* (2)	*inornata*	Densmore et al. (1989a); Dessauer and Cole (1989)
sonorae	3N	*burti* (2)	*inornata*	Densmore et al. (1989a); Dessauer and Cole (1989)
uniparens	3N	*burti*	*inornata* (2)	Densmore et al. (1989a); Dessauer and Cole (1989)
velox	3N	*inornata* (2)	*burti*	Densmore et al. (1989a); Dessauer and Cole (1989); Moritz et al. (1989c)

TABLE 3.1 *continued*

		Sexual parental species (subspecies)		
Parthenogenetic biotype	*Ploidy level*	*Males*	*Females*	*References*
opatae	3N	*burti* (2)	*inornata* (2)	Densmore et al. (1989a)
Kentropyx lizards				
borckiana	2N	*calcarata*	*striata*	Cole et al. (1995); Reeder et al. (2002)
Darevskia lizards				
armeniaca	2N	*valentini*	*mixta*	Moritz et al. (1992a)
bendimahiensis	2N	*valentini*	*raddei*	Fu et al. (2000)
dahli	2N	*portschinskii*	*mixta*	Moritz et al. (1992a)
rostombekovi	2N	*portschinskii*	*raddei*	Moritz et al. (1992a)
sapphirina	2N	*valentini*	*raddei*	Fu et al. (2000)
uzzelli[b]	2N	*valentini*	*raddei*	Fu et al. (2000)
uzzelli[b]	2N	*raddei*	*valentini*	Moritz et al. (1992a)
Gekkonidae geckos				
Heteronotia binoei	3N	races of *binoei*	races of *binoei*	Moritz (1993); Moritz and Heideman (1993)
Lepidodactylus lugubris	2N	undescribed species	*moestus*	Radtkey et al. (1995)

[a]For *Cnemidophorus* lizards in particular, the taxonomies and even the spellings of various species names have not been consistent in the scientific literature. The Latin names used here are mostly from Reeder and colleagues (2002), and in many cases they differ slightly from those employed in the original citations.

[b]Two molecular studies on this unisexual biotype reported the same parental species but opposite directions in the original cross(es). Assuming that neither study is in error (they involved different samples), the genesis hybridizations took place in both directions with respect to sex.

different and phylogenetically diverse pairs of hybridizing whiptail lizard species, seldom closely related, were involved in generating the parthenogenetic biotypes (see fig. 3.19). By contrast, Ilya Darevsky and colleagues have favored the phylogenetic constraint hypothesis for *Darevskia,* based on the observation that only a few sexual species in two distinct clades, or monophyletic groups, were the parents of all extant parthenogens (fig. 3.20), yet many other species of *Darevskia* hybridize without producing unisexual clones (table 3.2). Thus most, if not all, known *Darevskia* parthenogens arose from crosses between females of either *raddei* or *mixta* (two bisexual sister species in the *caucasica* clade) and males of *valentini*

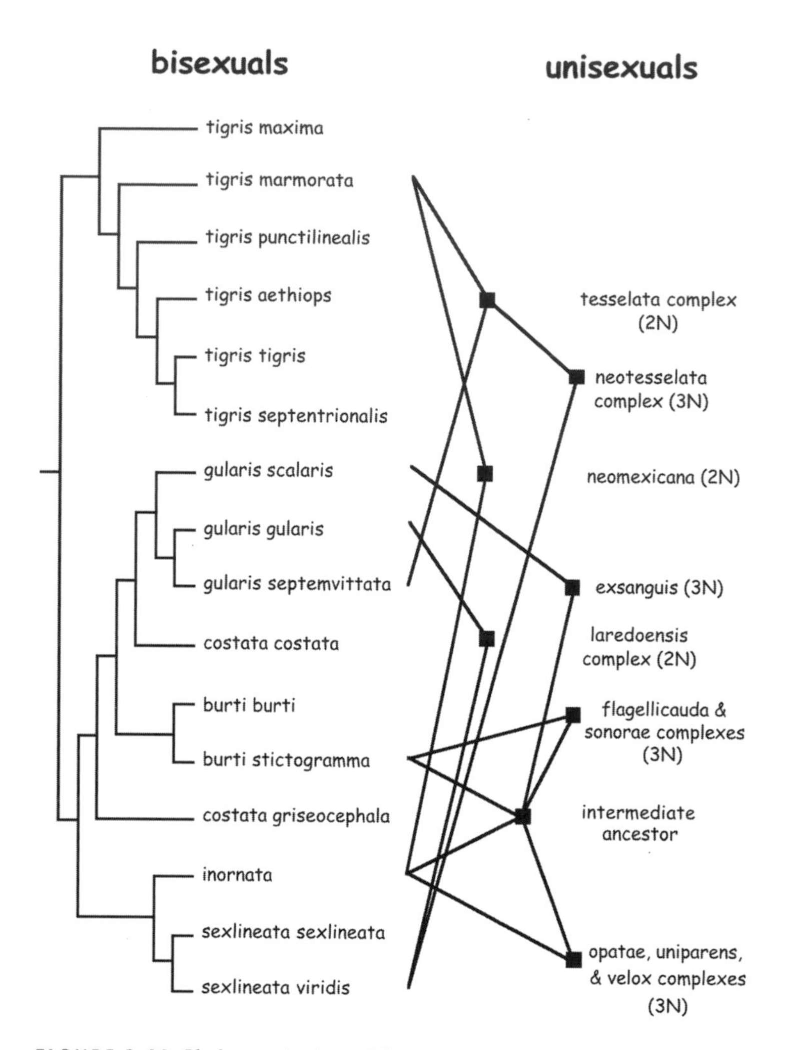

FIGURE 3.19 Phylogeny (estimated from mtDNA data) for representative sexual species of *Aspidoscelis* lizards and their unisexual derivatives (after Reeder et al., 2002). Note that the crosses yielding parthenogenetic biotypes were typically not between nearest genetic relatives among the sexuals.

or *portschinskii* (two bisexual near-sister species in the disparate *rudis* clade). The researchers speculated that a genetic incompatibility in hybrids between the sex chromosomes of these particular *Darevskia* species might account for the abnormal meiosis that yielded parthenogenetic rather than sexual lineages (Murphy et al., 2000).

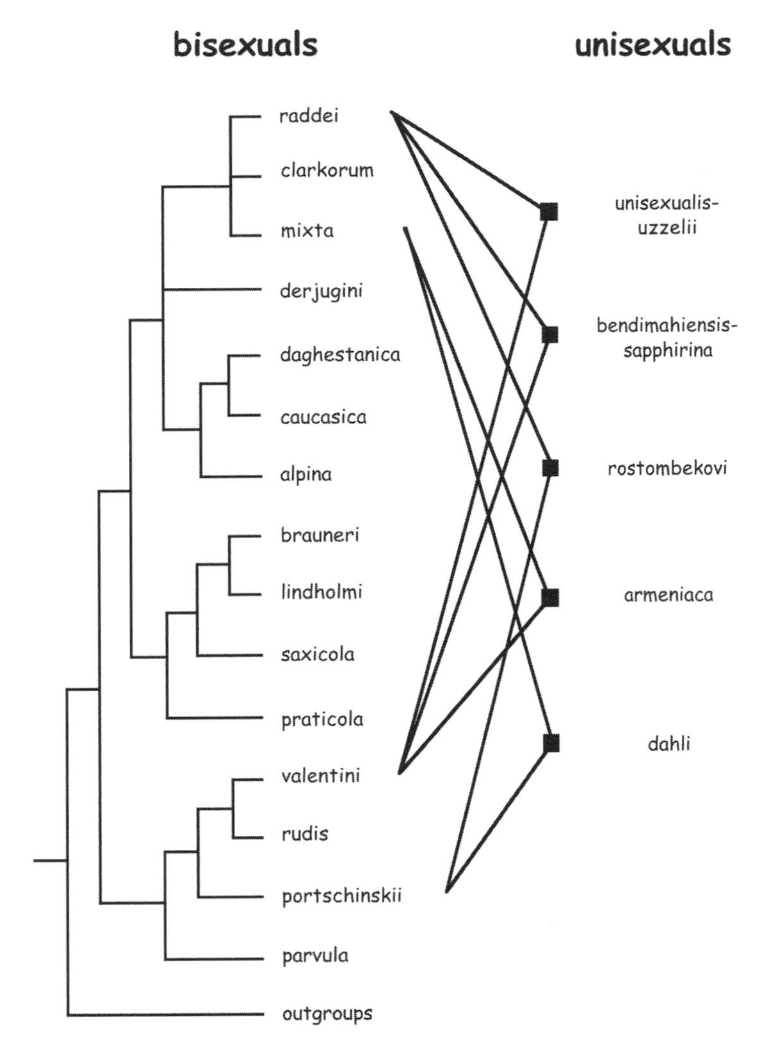

FIGURE 3.20 Phylogeny for 15 sexual species of *Darevskia* lizards and their diploid unisexual derivatives (after Murphy et al., 2000). Note that the crosses yielding extant parthenogenetic biotypes involve sexual species in two distant parts of the phylogenetic tree.

Triploid Geneses

For any triploid parthenogen, a question arises as to whether the initial hybridization event preceded or followed the production of an unreduced (diploid) egg by a diploid female. Under the primary-hybrid hypothesis (Schultz, 1969), a hybrid female produced an unreduced oocyte that subsequently was fertilized

TABLE 3.2 Pairs of bisexual *Darevskia* lizard species known to produce viable hybrids that are *not* parthenogens[a]

Closely related pairs	*More distantly related pairs*
alpina × *caucasica*	*alpina* × *brauneri*
caucasica × *daghestanica*	*caucasica* × *saxicola*
mixta × *derjugini*	*derjugini* × *parvula*
mixta × *alpina*	*rudis* × *clarkorum*
parvula × *rudis*	
saxicola × *brauneri*	

[a]Adapted from Murphy and colleagues (2000); phylogenetic relationships of these species are shown in figure 3.20. Similarly, for *Cnemidophorus* (*Aspidoscelis*) lizards, various bisexual species have been observed to hybridize without producing parthenogenetic progeny (see Moritz et al., 1992a).

by a haploid sperm (i.e., hybridization came first). Under the spontaneous-origin hypothesis (Cuellar, 1974, 1977), a nonhybrid female produced an unreduced oocyte that was then fertilized by sperm from a male of another sexual species (i.e., hybridization came second). These competing models entail different genetic consequences (fig. 3.21). The spontaneous-origin model predicts that the paired homospecific nuclear genomes in a parthenogenetic lineage derive from the sexual species that provided the maternal parent in the original hybridization, and thus should be coupled with maternally inherited mtDNA from that same parental species. By contrast, the most likely version of the primary-hybrid model predicts that the paired homospecific genomes in a parthenogenetic lineage derive from the sexual species that provided the male parent in the initial hybridization, and thus should not be paired with mtDNA from that same parent species.

Researchers have used cytonuclear genetic data to test these competing models in several parthenogenetic species of *Aspidoscelis* (*Cnemidophorus*). In at least eight of ten cases, results clearly support the primary-hybrid model and strongly reject the spontaneous-origin scenario (Densmore et al., 1989a; Moritz et al., 1989b). These findings support the conventional wisdom that interspecific hybridization is often the initial trigger for an atypical meiosis that sometimes eventuates in production of a triploid parthenogenetic lineage.

Number of Hybridization Events

Genetic evidence, either from direct molecular assays or from other kinds of genetic analysis such as histocompatibility responses in tissue grafts (e.g.,

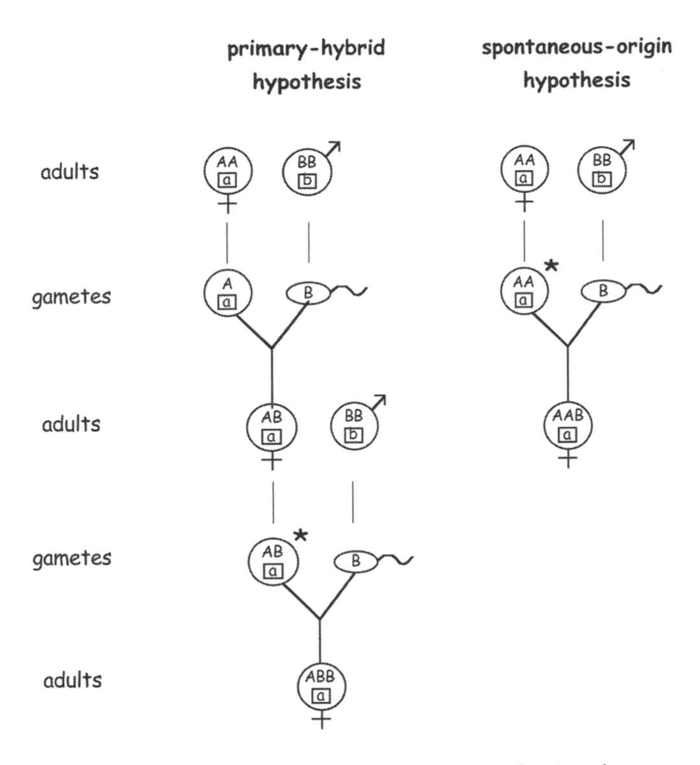

FIGURE 3.21 Schematic diagram of competing mechanisms for the origin of a triploid parthenogenetic lineage. Each uppercase letter represents one nuclear genome (A or B) from the respective parent species, and the lowercase letters in boxes similarly refer to the maternally inherited mtDNA genomes. Also shown are sperm (ovals) and eggs, the latter being unreduced where indicated by stars.

Hernandez-Gallegos et al., 1998; Abuhteba et al., 2000), can also be used to estimate the number of hybridizations that produced each parthenogenetic taxon. The task is complicated, however, by several factors discussed in box 3.4. For some unisexual biotypes, such as the diploid rock lizard *D. rostombekovi,* all extant members of a named species reportedly share the same multilocus nuclear genotype and therefore appear to be monoclonal, that is, of single hybrid origin (MacCulloch et al., 1997). Likewise, each of several triploid taxa of parthenogenetic reptiles seems to be monoclonal and monophyletic (Parker and Selander, 1976, 1984; Darevsky et al., 1984; Wynn et al., 1987; Dessauer and Cole, 1989), although more extensive genetic sampling may be desirable before definitive

BOX 3.4 Challenges in Deciphering the Number of Hybrid Geneses for a Unisexual Taxon

Deducing the precise number of hybrid events that produced a unisexual biotype is difficult for several reasons (Simon et al., 2003). First is the profound challenge of properly apportioning genetic differences observed among extant parthenogens to postformational mutations or to clonally propagated variation following independent hybrid origins (Butlin et al., 1999). Case-by-case judgment of the evidence is required. For example, unique alleles (mitochondrial or nuclear) that are rare and geographically localized in unisexuals probably arose through postformational mutations, whereas distinctive mtDNA genotypes that are common in unisexuals and shared with various extant sexual species probably derive from separate hybridizations. However, many outcomes are much harder to interpret because they are gray intermediates to these black-and-white extremes (e.g., Murphy et al., 1997; Fu et al., 1998).

A second challenge is to appreciate that a unique genesis for mtDNA in a unisexual taxon does not automatically equate to a single hybrid origin. Imagine that a single female hybridized with two or more males; the result would be multiple unisexual lineages that differ in nuclear genotypes yet belong to a single clade (monophyletic unit) in mtDNA. Thus mtDNA evidence is one-sided: multiple hybrid geneses are implied when a unisexual biotype displays distinctive matrilineal clades, but a single hybrid genesis is not proved by mtDNA monophyly. Indeed, several parthenogenetic vertebrate taxa that belong to a single mtDNA clade clearly arose through multiple hybridization events, as judged by the diversity of nuclear genotypes, involving closely related ancestral females (e.g., Moritz et al., 1989b; MacCulloch et al., 1995).

A third complication stems from taxonomy per se, as can be illustrated by the gecko *Heteronotia binoei*. This recognized biotype proved upon genetic inspection to consist of several cryptic sexual species as well as multiple parthenogenetic lineages independently derived from them (Moritz, 1993). So the nomenclature fails to capture the true biological diversity of this evolutionary complex of genetic lineages. In other cases, taxonomists have split the evolutionary assemblages far more finely, in which case a named unisexual taxon might indeed prove to be monophyletic. The point is that any face-value tally of the number of geneses for a particular parthenogenetic taxon will inevitably be influenced by how finely the unisexual biotype and its genetic relatives were taxonomically split to begin with.

conclusions are drawn. In some other cases, however, a recognized parthenogenetic species proved to be an ensemble of clonal lineages stemming from separate hybridization events. For example, the widespread Australian lizard *M. greyii* is comprised of at least three unisexual lineages of independent evolutionary origin, as well as at least three cryptic sexual species (Adams et al., 2003). Similar statements about the presence of multiple clonal origins apply to the geckos *H. binoei* (Moritz et al., 1989a) and *L. lugubris* (Moritz et al., 1993; Bolger and Case, 1994), and probably to some *Aspidoscelis* lizards, including diploid *A. laredoensis* (Walker et al., 1989) and the triploid complexes *A. cozumela, A. uniparens,* and *A. velox* (Fritts, 1969; Dessauer and Cole, 1989). In some cases, such as in the diploid parthenogen *A. cryptus* and also in *A. neomexicana,* genetic evidence indicates that extant clones originated from separate first-generation hybrid zygotes, but it is unclear whether those zygotes were from a single clutch or from distinct clutches involving different sets of parents (Parker and Selander, 1984; Cordes et al., 1990; Cole and Dessauer, 1993).

The broader question is as follows: Do parthenogenetic lineages arise commonly during reptilian evolution, or are they are extremely rare and special occurrences? Empirically, the answer lies somewhere in between. The evolutionary geneses are special in the sense that they require successful hybridizations between particular pairs of sexual species that are neither too close nor too distant genetically, and also by the fact that parthenogenetic biotypes (as tallied by any criterion) are vastly outnumbered by sexual species. On the other hand, parthenogens must originate rather routinely in the sense that a substantial number of reptilian genera and families include unisexual representatives, and many of these named biotypes are individually polyclonal because of multiple hybrid geneses. Furthermore, given that most parthenogenetic lineages are evolutionarily short-lived (see beyond), the number of evolutionary inaugurations for unisexual lineages during evolution must be severely underestimated by mere face-value tallies of clonal biotypes that have survived to the present time.

Genealogical Histories

Molecular genetic surveys have provided many insights into the genealogical histories of unisexual biotypes. In this regard, mtDNA molecules have been of special interest because the matrilineal history for any all-female taxon is in principle one and the same as the organismal phylogeny through which *all* loci—including nuclear genes—were transmitted (fig. 3.22, right). In any sexual species, by contrast, a matrilineal genealogy is only a minuscule fraction of a species' total genetic history, the vast majority of which involves nuclear loci that have trickled across a population's multigeneration pedigree along innumerable hereditary pathways, involving connections of sires as well as dams with their offspring (fig. 3.22, left).

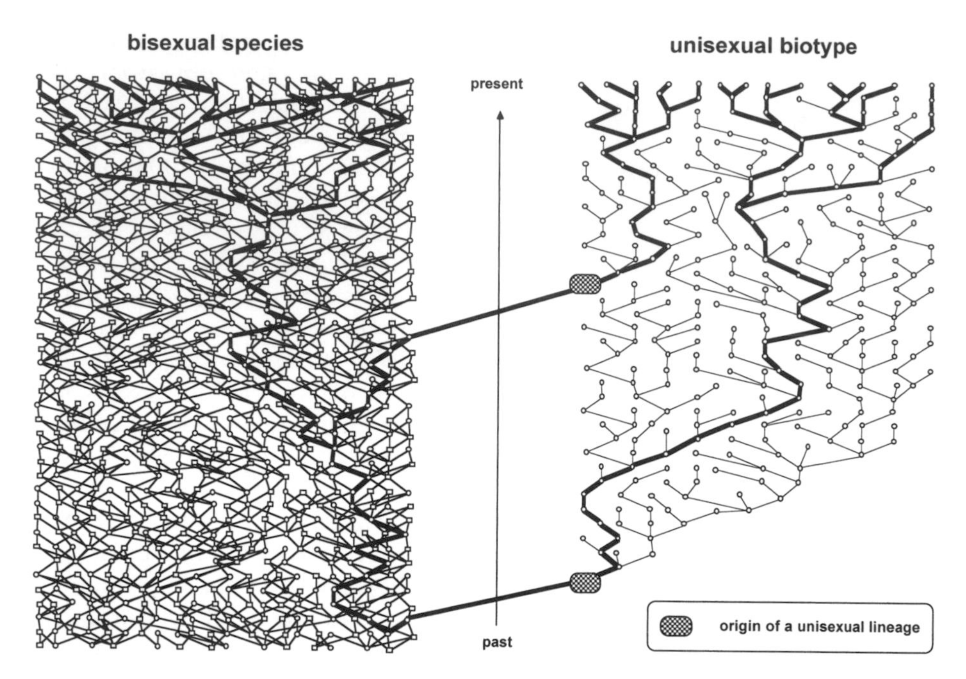

FIGURE 3.22 Sexual versus asexual pedigrees. *Left:* A hypothetical pedigree for a sexual species, with circles indicating females, squares indicating males, and lines connecting parents to progeny. *Right:* A hypothetical pedigree for an asexual (all-female) species. Note that each matrilineal genealogy, as indicated by thick lines, is only a tiny fraction of the organismal pedigree in any sexual species (*left*), but it is the entire pedigree for any asexual species (*right*).

Mitochondrial DNA phylogenies have been published for several parthenogenetic vertebrate taxa and their sexual relatives (review in Avise et al., 1992). In most cases, a particular unisexual taxon has proved to comprise only a single and relatively small branch in the broader matrilineal tree of the sexual species that provided its female parent in the original hybridization(s). This category of outcome is nicely illustrated by the parthenogenetic lizard *A. uniparens,* which as shown in figure 3.23 is a monophyletic entity nested within the matrilineal genealogy of *A. inornata,* its maternal sexual ancestor. Another way to characterize this pattern is to state that the ancestral sexual taxon is paraphyletic in matriarchal phylogeny with respect to its parthenogenetic derivative, meaning in this instance that *A. inornata* encompasses *some* but not *all* descendant lineages that stemmed from the ancestral node in its matrilineal tree.

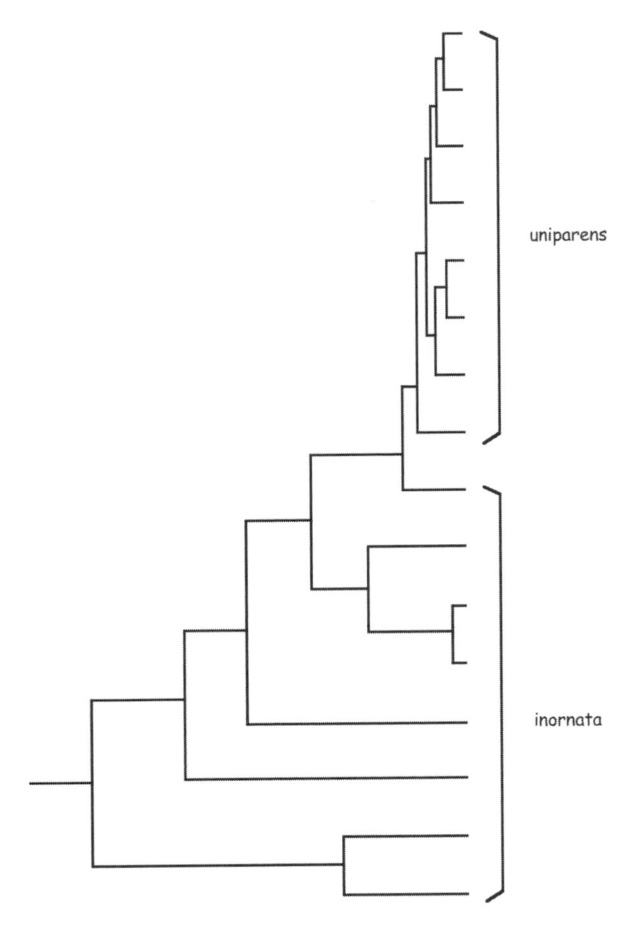

FIGURE 3.23 Matrilineal phylogeny (estimated from mtDNA data) for the sexual lizard species *Aspidoscelis inornata* and its parthenogenetic derivative *A. uniparens.*

The type of phylogenetic signature exemplified in figure 3.23 strongly suggests that extant individuals in a unisexual taxon trace their ancestry back to hybridization(s) involving just one female, or perhaps a few closely related females inhabiting a subset of the ancestral species' range. In the case of *A. uniparens,* the mtDNA genealogy and distributional data indicate that the maternal ancestor belonged to the *arizonae* geographic subspecies of *A. inornata* (Densmore et al., 1989a). The fact that *A. uniparens* comprises one twig on an outer branch

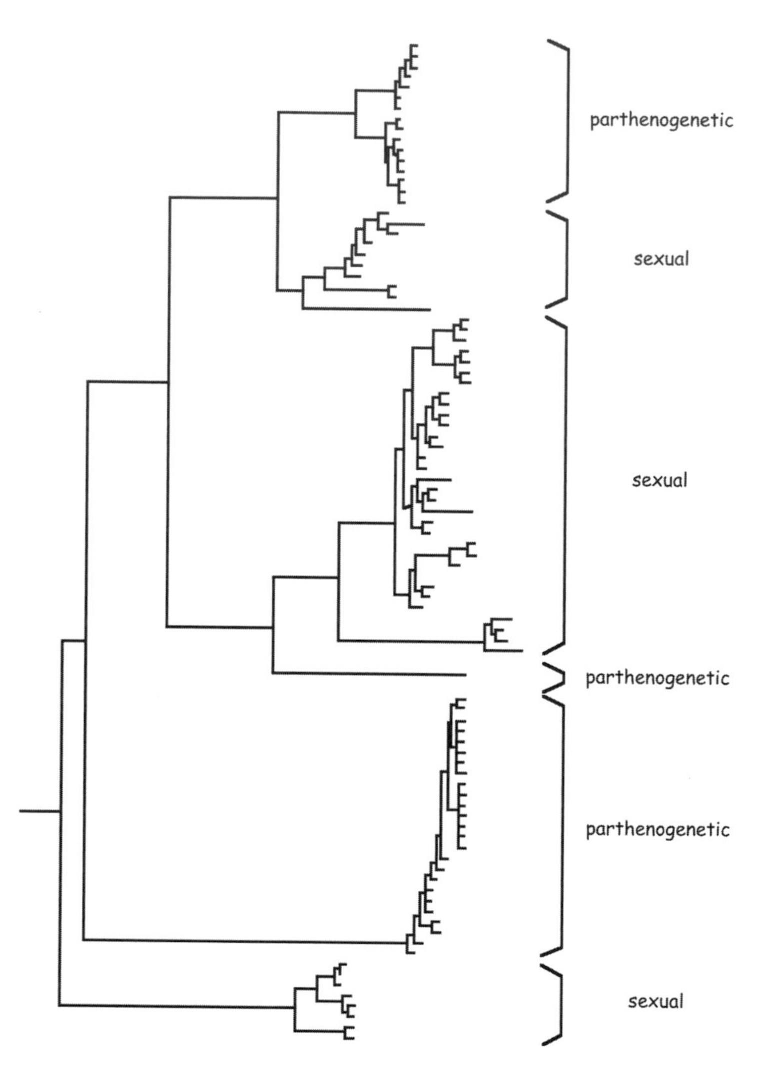

FIGURE 3.24 Matrilineal phylogeny (estimated from mtDNA data) for sexual and parthenogenetic forms of the Australian lizard *Menetia greyii* (after Adams et al., 2003).

of the much deeper matrilineal tree of *A. inornata* further implies that the parthenogenetic lineage is recent in origin relative to the matrilineal component of intraspecific history in its sexual ancestor.

At face value, a seemingly different category of phylogenetic outcome is illustrated by the Australian lizard *M. greyii*. As summarized in figure 3.24, this recognized taxon has proved to be a mixture of sexual and asexual populations, with parthenogenetic lineages showing evidence for least two or three separate hybrid origins within the matrilineal phylogeny of the sexuals. In other words, the parthenogenetic forms of *M. greyii* are probably polyphyletic in matrilineal ancestry. Note again, however, that such taxonomic and phylogenetic statements are inevitably intertwined. If, for example, each distinct unisexual lineage of *M. greyii* was instead considered a distinct taxonomic species, then each parthenogenetic taxon would be monophyletic, and the overall biological scenario would be more obviously analogous to that described above for the *A. uniparens/inornata* complex. (Or, conversely, if several currently recognized taxa of unisexual *Aspidoscelis* had been considered members of a single species, then that enlarged parthenogenetic taxon would by definition be polyphyletic, and the overall biological scenario would be more obviously analogous to that described for *M. greyii*.)

Such phylogenetic nuances can be further illustrated by reference to the matrilineal genealogy of *H. binoei*. This gecko taxon consists of several distinct sexual races (including SM6 and CA6), plus two parthenogenetic biotypes (3N1 and 3N2) that arose by repetitive hybridization events between SM6 and CA6. As shown in figure 3.25, individuals representing 3N1 constitute one shallow branch of the matrilineal tree for CA6, and individuals representing 3N2 constitute one shallow branch within the matrilineal tree for SM6. Regardless of the formal taxonomy adopted, the proper biological interpretation is that each parthenogenetic biotype is a subset of lineages nested well within the broader matrilineal genealogy of the sexual maternal ancestor.

Evolutionary Ages of Clones

With regard to the evolutionary longevities of parthenogens in nature, monophyletic clonal biotypes, not polyphyletic clonal taxa, are the relevant entities to be considered. Do particular parthenogenetic lineages typically persist for long, or are they evolutionarily short-lived? To address this question, two empirical approaches, both involving mtDNA, are popular (Avise et al., 1992): (1) assess the magnitude of sequence divergence between the parthenogen and the extant sexual relative to which it is genetically *closest;* and (2) assess the magnitude of sequence variation within the clonal lineage. In both approaches, the underlying rationale is that mtDNA sequence differences tend to accumulate over time. In several taxonomic groups of vertebrates, mtDNA molecules are known to evolve at a rate of approximately 1% sequence change per million years per lineage (or 2% sequence divergence per million years between lineage pairs). Although reservations apply to the precision and universality of any calibration of a "molecular clock," the genetic findings do permit at least ballpark estimates of evolutionary ages for clonal lineages vis-à-vis those of their cognate sexual species.

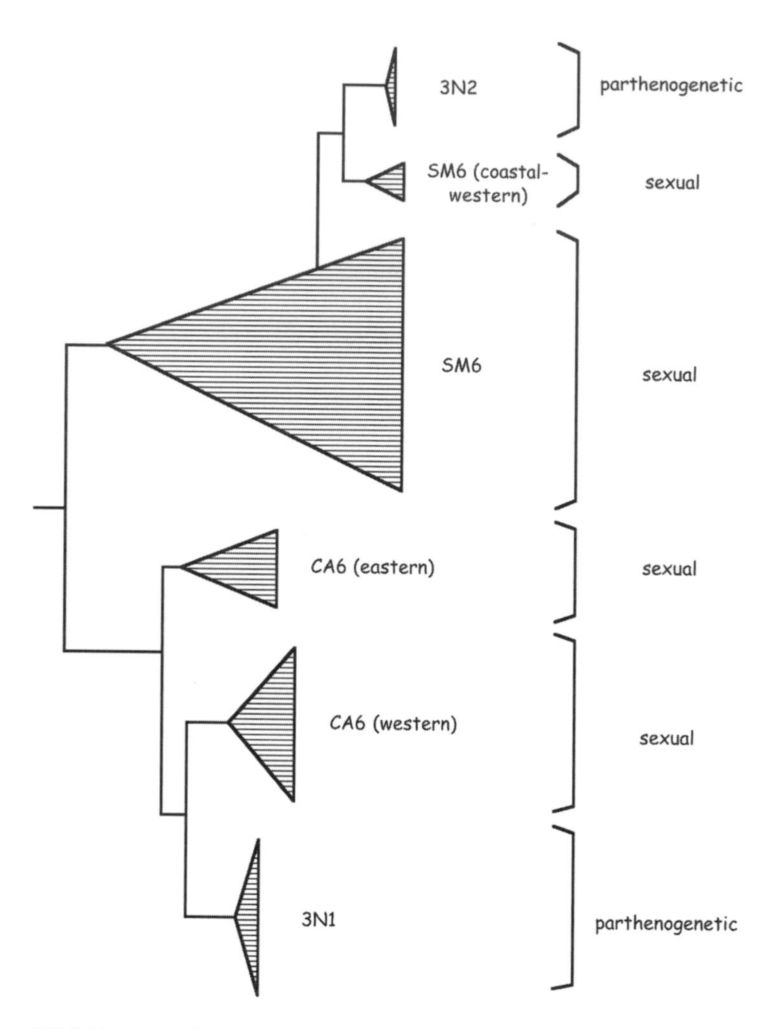

FIGURE 3.25 Matrilineal phylogeny (estimated from mtDNA data) for sexual and parthenogenetic forms of the Australian gecko *Heteronotia binoei* (after Kearney et al., 2006).

The first approach—assessing genetic divergence between an asexual biotype and its sexual kin—can set a lower (most recent) bound on the estimated age of a parthenogenetic lineage. But it suffers from the potentially serious complication that the closest sexual taxon may have gone extinct or otherwise remained unsampled, in which case the parthenogen could appear at face value to be much more ancient than it truly is. The second approach—quantifying genetic diversity within an asexual biotype—avoids this difficulty, but is subject to two other complications: low genetic variation within the parthenogen might be due to postgenesis population bottlenecks, rather than a recent clade origin; or, conversely, high genetic variation in the parthenogen might be due to unrecognized polyphyly, rather than an ancient clade origin.

In practice, these caveats have proved to be somewhat moot because, empirically, mtDNA sequences within most extant unisexual clades show both limited variation (fig. 3.26) *and* close similarity to genotypes still present in at least some extant populations of their sexual cognates. For example, nucleotide diversity

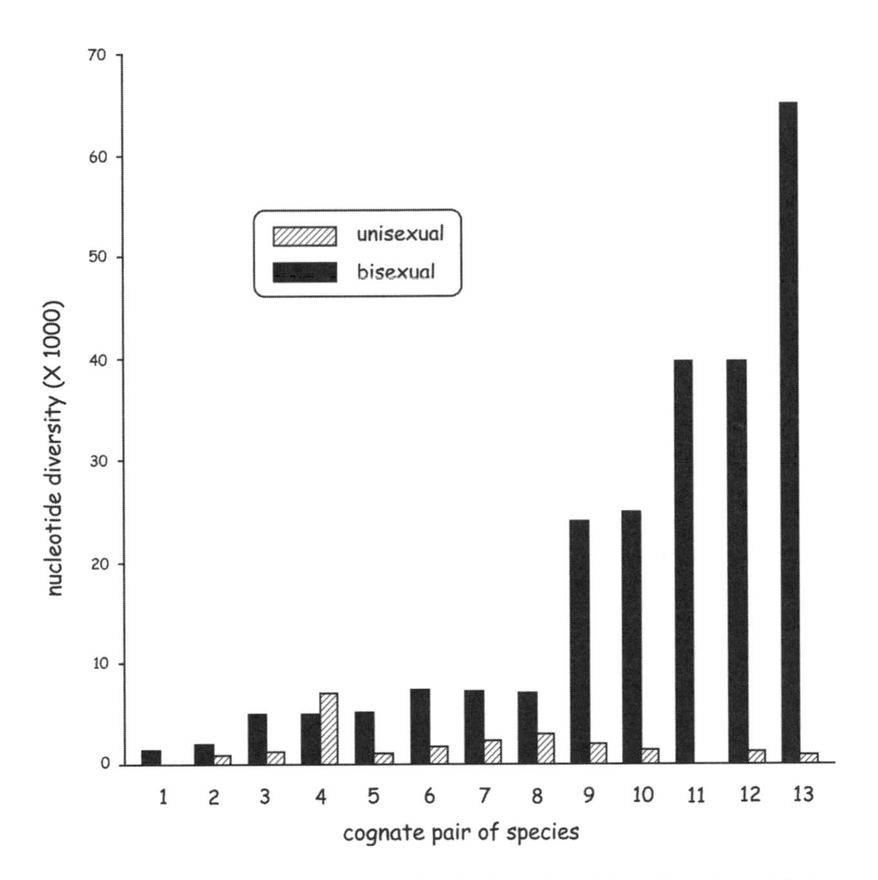

FIGURE 3.26 Mitochondrial DNA nucleotide diversities in bisexual species and their respective unisexual derivatives, arranged in rank order from left to right by the magnitude of genetic variation within the bisexual taxa (after Avise et al., 1992). About half of the comparisons in the graph involve parthenogenetic lineages and the other half involve gynogenetic or hybridogenetic lineages.

(mean DNA sequence divergence among specimens) within the unisexual taxon *A. uniparens* (0.2%) is an order of magnitude lower than nucleotide diversity within its sexual parent species *A. inornata* (2.4%), and all mtDNA sequences in the parthenogen are closely related to a subset of those in *A. inornata* (Densmore et al., 1989a). Both of these features of the data are also evident in the composite matrilineal phylogeny for *A. uniparens* and *A. inornata* (see fig. 3.23).

Qualitatively similar outcomes apply to nearly all unisexual vertebrate lineages that have been surveyed to date for mtDNA, thus demonstrating that most extant parthenogenetic clones and clades arrived recently on the evolutionary stage. For biological and technical reasons, confidence limits on the estimated sidereal ages of modern unisexual lineages are wide, often extending from zero (the present) to several hundred thousand years ago. Representative quantitative examples can be found regarding the ages of *H. binoei* (Kearney et al., 2006) and *D. rostombekovi* (Darevsky et al., 1986; Fu et al., 2000).

Despite such uncertainties about absolute dates, unisexual lineages in vertebrates are characteristically young *compared to* extant sexual species, which by the same kinds of mtDNA evidence typically originated (split phylogenetically from an extant sister species) about 0.5–3.0 million years ago (Klicka and Zink, 1997; Avise et al., 1998). Nevertheless, any conclusion that parthenogenetic lineages are invariably short-lived must be tempered by the possibility, not yet falsified, that at least a few parthenogenetic invertebrate clades may be tens or even hundreds of millions of years old (see box 3.2 and citations therein).

The young evolutionary ages of extant parthenogens do not necessarily imply, of course, that the unisexual phenomenon itself began only recently. Parthenogenetic lineages probably have been arising and then quickly going extinct for a long time. Tod Reeder and colleagues suggest that the ability to produce parthenogenetic hybrids may extend back throughout the >200 million year history of the squamates (2002).

Comparative Ecology and Natural History

Despite their relatively short evolutionary durations, some parthenogenetic biotypes can be highly successful over ecological timescales as judged by their wide geographic distributions, current abundances, and/or particular adaptive features (Kearney, 2005). Any short-term ecological successes as well as long-term evolutionary failures of parthenogenetic biotypes are both of academic interest, because they should help reveal how the costs and benefits of asexual versus sexual reproduction might play out on different temporal scales (Case and Taper, 1986).

At least three salient genetic features of potential ecological significance distinguish all parthenogenetic biotypes from sexual species. The first feature—extremely high heterozygosity (within-individual genetic variation) in the nuclear genome—stems from the hybrid origin of each clonal lineage. If heterosis (a higher mean fitness of heterozygotes than homozygotes) is a common phenomenon in unisexual taxa (see Hotz et al., 1999, for one such example), then elevated heterozygosity could substantially promote the survival and reproduction of parthenogenetic animals relative to sexuals. The second feature is "fertilization assurance" afforded by the capacity of a parthenogenetic individual to reproduce without a mate. This ability could pay large fitness dividends, especially in ecological circumstances where mate acquisition might otherwise be difficult—for example, in low-density populations, or in species where individuals have proclivities to disperse widely and colonize distant sites. The third special feature of parthenogenetic lineages is an effective absence of genetic recombination, which in theory could have opposing ecological ramifications: an opportunity for individuals to perpetuate genotypes that have been selection-tested or proven to work in a particular ecological setting, but also the potential liability of producing broods that lack genetic variety apart from de novo mutations. Also of potential negative import for any parthenogenetic lineage is its diminished ability to receive, via mating, undefiled template nucleic acid molecules for the potential repair of nuclear DNA damages, and presumably a greater susceptibility to Muller's ratchet.

Natural selection continually tallies all such costs and benefits on the fitness ledgers. Some of the costs, in particular, may be delayed or cumulative within a parthenogenetic clone, and thus may have negative biological ramifications that tend to increase with the evolutionary age of the unisexual clade. Such time-delayed costs might include an evolutionary accumulation of deleterious mutations, periodic or cumulative changes in the physical environment that fall beyond the adaptive scope of a parthenogenetic lineage, or a constrained capacity to coevolve apace with challenging biotic agents such as pathogens and parasites (see beyond). On the other hand, an ancient (as opposed to recent) parthenogenetic lineage may have accumulated at least some adaptively relevant genetic variation through either postformational mutations (West et al., 1999) or perhaps occasional gene conversions or other recombination events that have been implicated in a few studies (see, e.g., Moritz, 1993; Kan et al., 2000). These could potentially give the unisexual lineage a collectively broader scope for withstanding future environmental challenges.

Conventional wisdom long held that parthenogenetic taxa are competitively inferior to sexual species and that each unisexual lineage is restricted to a narrow ecological niche (Cuellar, 1977). This notion of parthenogens as fugitives or weeds (Wright and Lowe, 1968) was bolstered by the fact that these unisexuals in effect have "fertilization assurance" (Baker, 1965) that may enhance their ability (relative to sexuals) to colonize new areas. Although such characterizations may have elements of truth, the abundances and distributions of parthenogens show varied patterns and undoubtedly can be affected by many ecological and evolutionary factors, as described next.

Geographic Distribution and Abundance

The geographic distributions of parthenogenetic biotypes range from local to widespread (as is also true for their sexual relatives). Unisexual *Aspidoscelis* biotypes fall into the former category because each taxon is confined to a relatively small region in the southwestern deserts of North America (fig. 3.27). Similarly, each unisexual taxon of *Darevskia* lizard has a quite narrow geographic range in the Caucasus region (fig. 3.28). Examples of parthenogens with far wider distributions are the two clades of *H. binoei* in western and central Australia (fig. 3.29), and two other geckos (*L. lugubris* and *N. pelagicus*) whose unisexual biotypes can be found on numerous islands in the Indo-Pacific region (fig. 3.30).

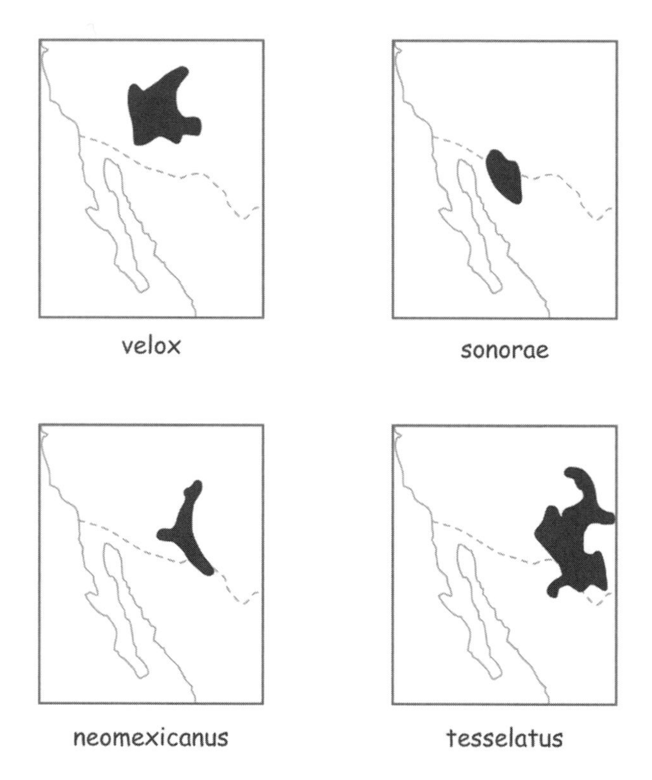

FIGURE 3.27 Geographic ranges of some representative parthenogenetic *Aspidoscelis* (*Cnemidophorus*) species (after Smith and Brodie, 1982).

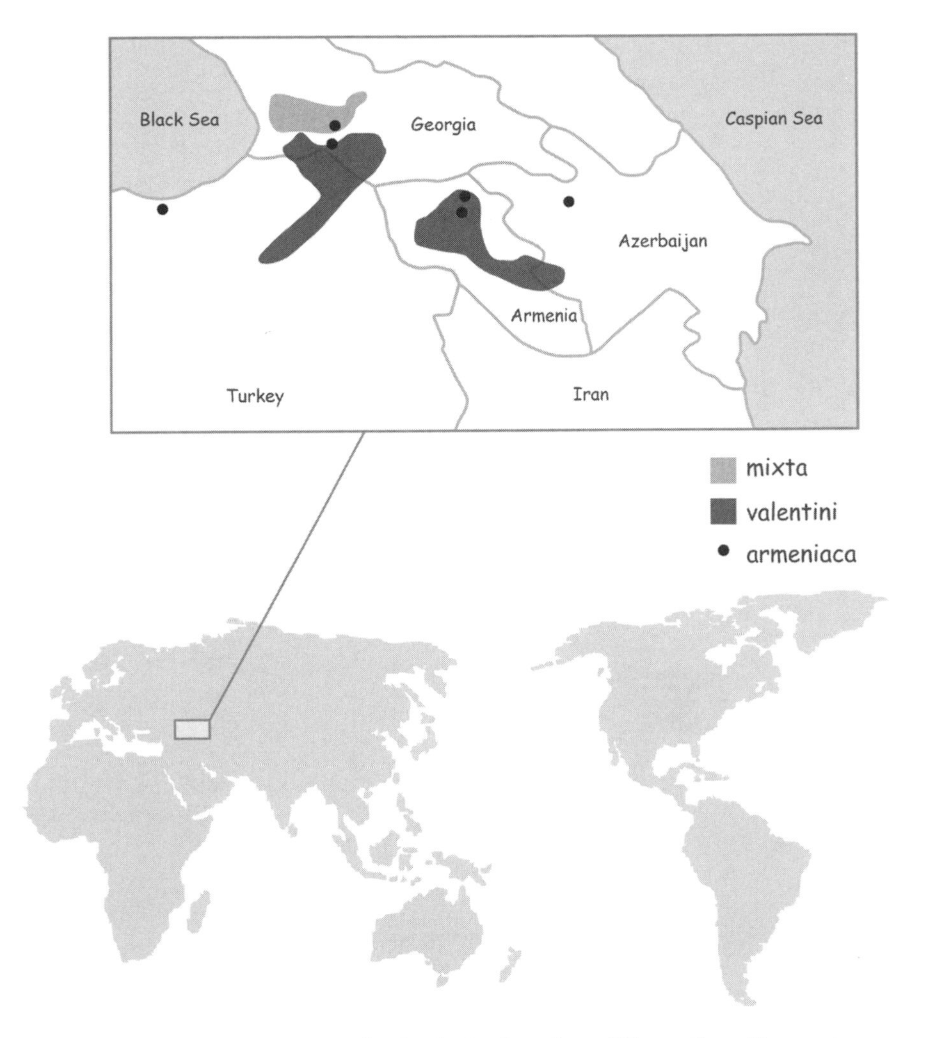

FIGURE 3.28 Some known geographic locales for the unisexual biotype *Darevskia armeniaca,* and the current geographic distributions of the two sexual species that give rise to it (after Moritz et al., 1992a).

From inspections of such range maps, three key points emerge.

1. The full distribution of a parthenogen is seldom confined within the geographic boundaries of its sexual parent taxa. This makes sense, mechanistically, because parthenogens (unlike gynogens and hybridogens, to be discussed in chapter 4) do not require the sexual services of a gonochoristic

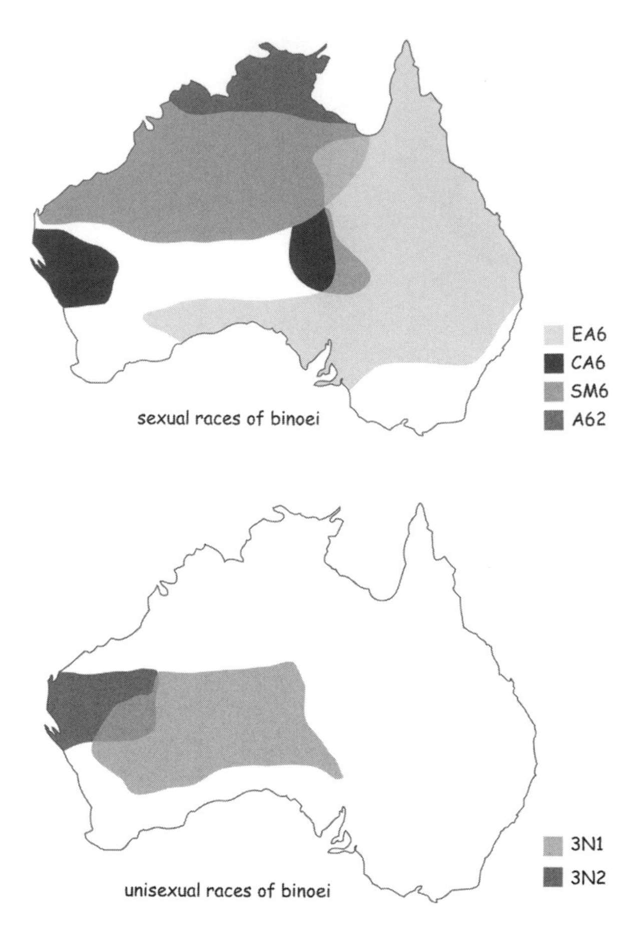

FIGURE 3.29 Range maps for various sexual (*top*) and asexual (*bottom*) genetic races of *Heteronotia binoei* (after Moritz, 1993).

species. Such range extensions also suggest that parthenogens can sometimes exploit ecological opportunities that have not been capitalized upon by their sexual progenitors.

2. The geographic range of a parthenogen often overlaps extensively with those of related sexual species. Such sympatry is common in *Aspidoscelis, Darevskia, Heteronotia,* and others, and it implies that unisexuals are not necessarily displaced by sexuals, at least at these coarse spatial scales (but see Petren et al., 1993, for a probable case of a sexual gecko having invaded and displaced an asexual native).

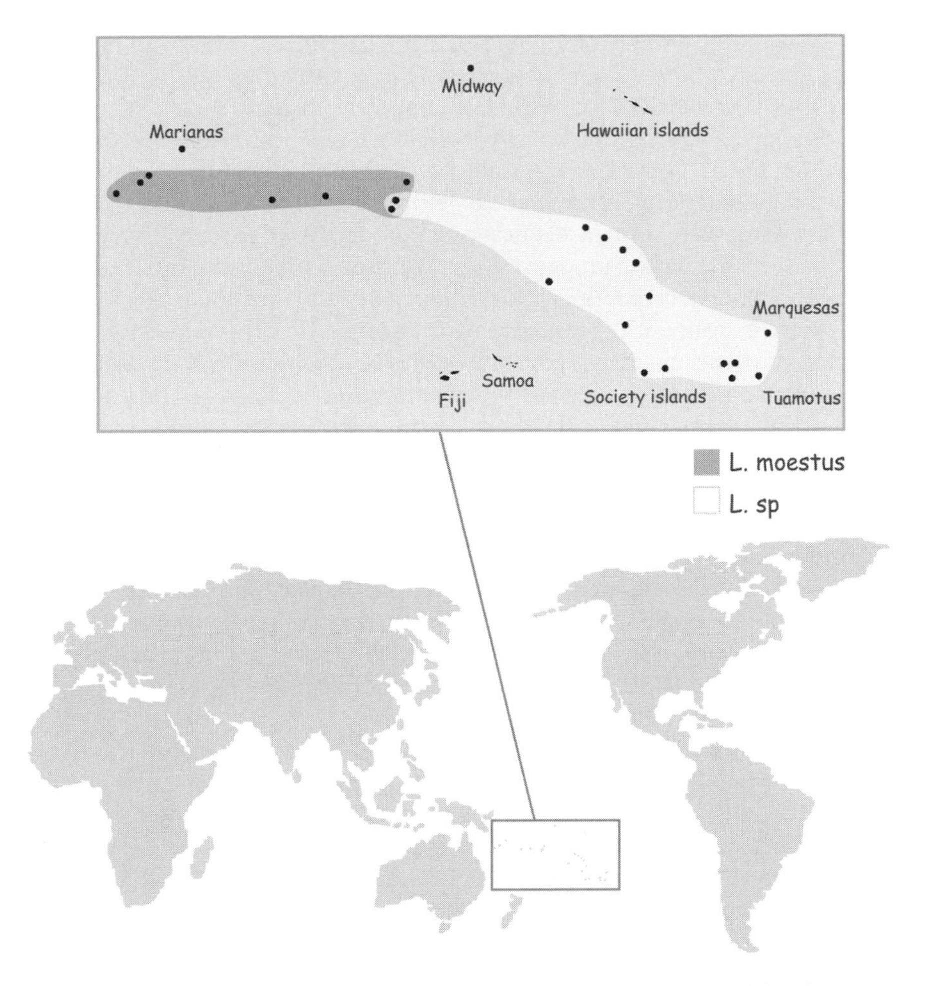

FIGURE 3.30 Current geographic distributions of the two sexual species (*Lepidodactylus moestus* and *L. sp.* undescribed) that give rise to unisexual *L. lugubris* (after Radtkey et al., 1995). The range of *L. lugubris* also includes many additional islands in the Indian and Pacific oceans.

3. Some parthenogenetic clades have broad distributions, so either the lineage is ancient, or it colonized the area recently via rather rapid dispersal. Genetic and ecological reasoning typically suggest that the latter explanation is more likely.

For example, the 3N1 unisexual clade of *H. binoei* shows very low mtDNA nucleotide diversity—0.1%, compared to 3.7% in its sexual progenitor CA6—implying a single geographic point of origin for this clade despite its current vast range that extends nearly 2,000 km across western and central Australia (see fig. 3.29). From a genetic point of view, if we provisionally assume a standard mtDNA clock, then a random pair of lineages differing by 0.1% in nucleotide sequence may have separated about 50,000 years ago, and oldest separations within the unisexual clade would likely have occurred about two to four times earlier (i.e., 100,000–200,000 years ago). From an ecological point of view, if we conservatively assume that each gecko disperses, on average, 0.1 km from its birth site (not necessarily in the same compass heading), then a distance of 2,000 km could certainly have been traversed within a comparable time frame of tens of thousands to at most a few hundred thousand animal generations. Thus, despite 3N1's wide distribution on the Australian continent, both the genetics and ecology can be interpreted as supporting a scenario of fairly recent origination and rapid spread of this unisexual clade.

Unisexual geckos on oceanic islands give an even clearer signal of recent origin and rapid colonization. For example, the unisexual biotype *L. lugubris* occurs on numerous islands in the Indian and Pacific oceans (see fig. 3.30), but it displays low mtDNA nucleotide diversity and close sequence similarity to its maternal sexual parent species, *L. moestus*, suggesting a young evolutionary age. The animals probably colonized many of these otherwise inaccessible sites via chance hitchhiking aboard the boats of Polynesians and Melanesians, who peopled the Indo-Pacific during the last three millennia (Zug, 1991; Moritz et al., 1993). For the parthenogenetic geckos, the capacity of each individual to reproduce without a mate undoubtedly facilitated this recent colonization process.

The abundance of individuals in some clonal taxa is further testimony to a parthenogen's capacity for at least short-term ecological success. For example, where they occur sympatrically in French Polynesia, asexual geckos (*L. lugubris*) outnumber sexual geckos in most habitat types (Hanley et al., 1994), and parthenogenetic *Darevskia* lizards are typically more numerous than their sexual counterparts on their broadly overlapping habitats in the Caucasus Mountains (Darevsky et al., 1985). On the Australian continent, unisexual forms of another gecko—*H. binoei*—are common within a range that overlaps only marginally with the sexual races (Moritz et al., 1989b, 1990). Joan Whittier and colleagues (1994) interpreted this distributional pattern to suggest that the parthenogens tend to displace the sexuals, ecologically, where they might otherwise co-occur; whereas Michael Kearney (2005) interpreted the pattern to suggest that the asexual hybrids can better colonize and exploit newly opened environments in the Australian deserts (but see also Kearney, 2006; Lundmark, 2006). Many additional parthenogenetic biotypes are likewise common and successful today, at least locally. Occasionally, related unisexual biotypes also overlap microspatially, and perhaps ecologically, as in some taxonomic species of parthenogenetic *Darevskia* lizard that have even been observed basking on the same rocks (Uzzell and Darevsky, 1975).

Adaptive Phenotypes

Unisexual lineages typically differ from one another and from their sexual ancestors in particular morphological, physiological, and ecological features (Darevsky, 1966; Parker et al., 1989; Kearney and Shine, 2004a, 2004b, 2005). For example, most of the parthenogenetic species of *Aspidoscelis* (*Cnemidophorus*) lizard have distinct and heritable skin-pigmentation patterns (Dessauer and Cole, 1989), and they also sometimes occupy different habitats (e.g., pinyon-juniper forests for *A. velox* versus desert-grasslands for *A. uniparens*). In *Lepidodactylus* geckos, sympatric parthenogenetic lineages have been found to differ in several ecological attributes, including the animals' thermal preferences, microspatial arrangements, and activity patterns (Bolger and Case, 1994). Such findings are not unduly surprising because the genotype of each parthenogenetic lineage is a unique amalgam of genes from two or more species that in effect was clonally "frozen" when the hybrid biotype arose (Vrijenhoek, 1984a).

More interesting is whether the phenotypes of unisexual biotypes differ consistently from those of sexual species in ways that impact ecological success. For example, perhaps the high heterozygosity *within* each parthenogen affords the animals a wider ecological tolerance, either because of heterozygote advantage per se (the heterosis hypothesis; Schultz, 1977; Bulger and Schultz, 1979) or because between-lineage selection has winnowed surviving lineages to those that happened to have broader niches (the general-purpose-genotype hypothesis; Parker, 1979; Lynch, 1984b). Alternatively, perhaps low genetic diversity *among* clonal individuals limits the collective niche width of a parthenogenetic biotype compared to that of a sexual species (as was shown to be true with respect to the diets of asexual versus sexual *Aspidoscelis;* Case, 1990). In addressing such possibilities, a clear distinction should be drawn between monophyletic and polyphyletic parthenogenetic taxa (because, e.g., composite phenotypic diversity in a monoclonal biotype might be much narrower than that in a polyclonal assemblage; Parker et al., 1989; Weeks, 1995; Schlosser et al., 1997).

Moritz and colleagues (1991) compared the incidence of blood-feeding mites (genus *Geckobia*) on parthenogenetic versus sexual individuals of *H. binoei.* The concept being tested—one version of the red queen hypothesis—was that asexual lineages, lacking genetic recombination, have been less able than sexual lineages to coevolve apace with their attacking parasites (Lively et al., 1990). Parasite loads indeed proved to be higher in the unisexual lineages (fig. 3.31). It was not determined, however, whether the mites actually impact survivorship or reproduction in these geckos.

The parasite data summarized in figure 3.31 are stunning, but their generality has been questioned. From similar kinds of observations on other geckos (in the genus *Lepidodactylus*), Kathryn Hanley and colleagues (1995) reported that asexuals have *lower* infestations of mites than sexuals sharing the same habitat. These authors suggested that the asexual geckos' low susceptibility to parasites

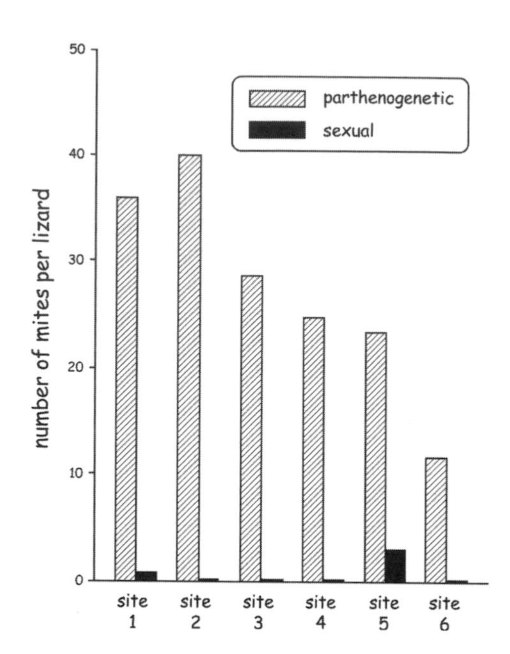

FIGURE 3.31 Levels of mite infection in parthenogenetic and sexual individuals of *H. binoei* at each of six geographic sites where these geckos co-occur (after Moritz, 1993).

might be due to genetic resistance stemming from the elevated heterozygosity that traces back to the parthenogen's hybrid origin. A similar hypothesis was advanced for why asexual lineages of gecko can persist for thousands of generations in the demonstrated absence of recombination in pathogen-resistance genes of the major histocompatibility complex (Radtkey et al., 1996).

Another example of contrasting outcomes has been reported with respect to physiological adaptations in parthenogenetic versus sexual lizards. For *H. binoei* geckos, Kearney and colleagues (2005) found that parthenogenetic animals were capable of greater aerobic activity than their sexual counterparts, but Alistair Cullum (1997) reported the opposite outcome for parthenogenetic and sexual forms of *Aspidoscelis* (*Cnemidophorus*) lizards.

Several studies have reported that unisexual taxa, in various invertebrate groups as well as some vertebrates, have lower per-individual fecundities or reproductive rates than their sexual counterparts (reviewed in Lynch, 1984b, p. 268). For example, parthenogenetic *Heteronotia* geckos reared in the laboratory displayed about 30% lower fecundity than their sexual progenitors, irrespective of body size (Kearney and Shine, 2005). If this reproductive tendency proves to be general, it could be another factor contributing to the limited long-term success of parthenogens.

More empirical studies of these types are needed before any sweeping generalities might be drawn about the comparative ecologies of vertebrate parthenogens and sexual reproducers. In the meantime, three conclusions already seem undeniable: some parthenogenetic biotypes can be highly successful in the short or moderate term; such ecological success does not continue indefinitely, as judged by the fact that no parthenogenetic lineage in vertebrates is demonstrably ancient; and, given the heterogeneity of outcomes already recorded, ecological and genetic factors that are idiosyncratic to particular taxa (in addition to broader biological considerations) will be needed to explain the overall distribution of parthenogenesis among the vertebrates.

CHAPTER FOUR

Reproduction by the Semichaste: Gynogenesis, Hybridogenesis, and Kleptogenesis

Gynogenesis ("the origin of females") and hybridogenesis ("the origin of hybrids") are processes similar to parthenogenesis, but with added dashes of sex. Females in the clonal or quasi-clonal taxa that utilize these reproductive modes are not virgins; rather, each mates with a male from a foreign species and utilizes his sperm. Remarkably, however, these males typically make no lasting genetic contribution to future generations of the unisexual (all-female) lineage. They are the duped victims of sexual parasitism. How such peculiar reproductive and genetic events transpire, and which vertebrate organisms engage in these and related sperm-dependent forms of clonality and quasi-clonality, are the subjects of this chapter.

Gynogenesis (in its most basic form) is diagrammed in figure 4.1. The process is like parthenogenesis except that a female's genetically unreduced egg does not begin to divide and multiply until it is physically poked or pricked by a sperm cell. The sperm does not fertilize the egg, but merely stimulates it to begin dividing, thereby leading to the development of a new individual. Via this "pseudogamous" process, the resulting gynogenetic daughter is genetically identical to her mother, and usually carries no genes from her sexually parasitized "father."

Hybridogenesis (in its most basic form) is diagrammed in figure 4.2. This sperm-dependent operation differs from gynogenesis in three key regards. Specifically, each egg produced by a hybridogenetic female is "reduced" in chromosome constitution (e.g., it is haploid rather than diploid); sperm from a male of the foreign sexual species actually enters and fertilizes that egg, thereby reestablishing the unreduced ploidy condition in the resulting offspring; and then later, during gametogenesis in a mature daughter, the paternal chromosomes

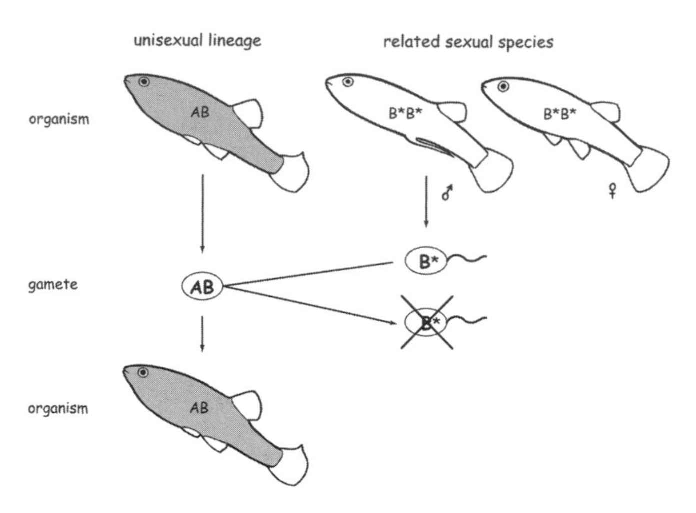

FIGURE 4.1 A simple summary of gynogenesis. A and B refer to distinct nuclear genomes from two sexual species that hybridized to initiate the gynogenetic lineage.

are physically excluded during an abnormal meiosis in which the chromosome sets from the mother and father also typically fail to recombine. The net consequences of this peculiar process are as follows: a sexually parasitized male is a biological father but he cannot become a genetic grandfather; only maternal genes and chromosomes are perpetuated across generations of the unisexual biotype; and genetic recombination is generally absent despite the biological involvement of two separate parents in each offspring's production. Hybridogenesis (even in its simplest form) thus has elements of both sexuality and asexuality, and it is sometimes referred to as a hemiclonal system.

The Cast of Players

Fish and amphibians provide all of the known vertebrate examples in which all-female clones or hemiclones reproduce in nature by constitutive gynogenesis or hybridogenesis. About 50 named species of sperm-dependent unisexual vertebrates have been described, representing about a dozen genera in seven taxonomic families. All of these taxa also contain sexual species from which the unisexuals arose via interspecific hybridization. The families that contain clonal or hemiclonal biotypes will be briefly introduced in the following sections, and one gynogenetic or hybridogenetic species representing each genus will be illustrated.

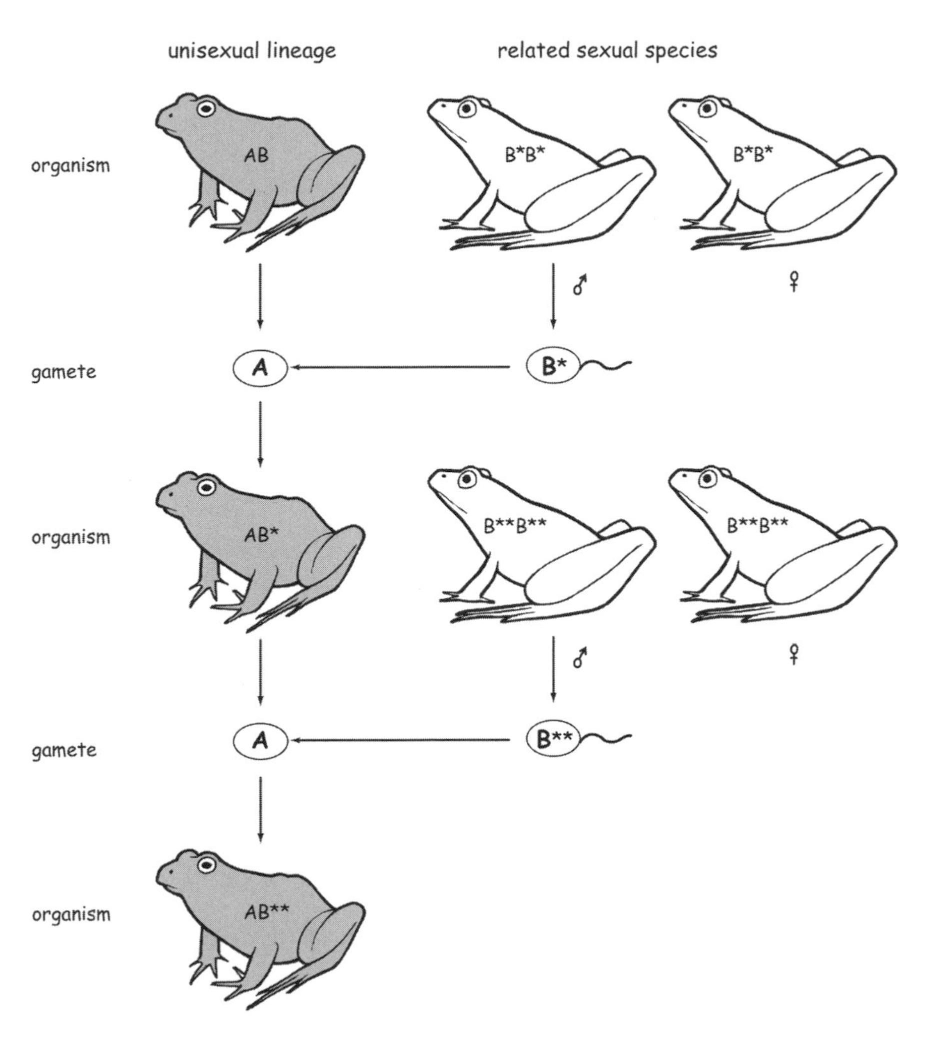

FIGURE 4.2 A simple summary of typical hybridogenesis. A and B refer to distinct nuclear genomes from two sexual species that hybridized to initiate the hybridogenetic lineage.

Poeciliidae (Live-Bearing Fishes)

Members of Poeciliidae have internal fertilization, and females gestate the embryos before giving birth to broods of live young. More than 130 extant species in about 20 genera inhabit freshwater and brackish environments of North and South America. Two genera (*Poecilia* and *Poeciliopsis*) have unisexual representatives, in each case including both diploid and triploid unisexual forms.

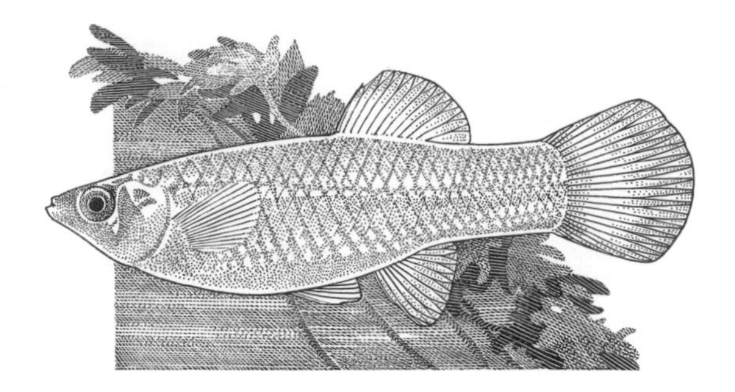

FIGURE 4.3 *Poecilia formosa,* the Amazon molly.

Poecilia (Formerly Mollienesia) formosa

The Amazon molly (fig. 4.3) is a mostly diploid gynogen native to freshwater habitats in northeastern Mexico and southeastern Texas (Kallman, 1962a; Turner et al., 1983). Discovered in 1932 by Carl and Laura Hubbs (box 4.1), it was the first clonal vertebrate conclusively known to science, and it is also of historical importance for being perhaps the first clonal animal on which tissue transplantation experiments were conducted (box 4.2). Later, molecular genetic data revealed that this all-female biotype arose quite recently in evolution via hybridization between two sexual species: *P. mexicana* as the original female parent species, and a close relative of extant *P. latipinna* as the original male parent (Avise et al., 1991; Schartl et al., 1995b). The gynogens continue to use these parental species, as well as *P. latipunctata,* as sexual hosts (Niemeitz et al., 2002; Schlupp et al., 2002). Also present in this genetic assemblage are triploid biotypes (Schultz and Kallman, 1968; Rasch and Balsano, 1989; Lampert et al., 2005) that appear to have had more than one evolutionary origin (Schories et al., 2007), and even some somatic mosaics (Lamatsch et al., 2002) in which the reproductive mode is also basically gynogenetic. Rare "leakage" of some paternal genomic material into "gynogenetic" lineages has also been reported (Kallman, 1964).

Poeciliopsis monacha-lucida

Poeciliopsis monacha-lucida, an all-female species (fig. 4.4), is a diploid hybridogen native to desert arroyos in northwestern Mexico. As its hyphenated name implies, it arose via crosses between the sexual species *Po. monacha* (the female parent) and *Po. lucida* (the male parent). Also in this taxonomic group are gynogenetic triploid biotypes, notably *Po. 2 monacha-lucida,* which carries two nuclear genomes from *monacha* and one from *lucida,* and *Po. monacha-2 lucida,* with two genomes from *lucida* and one from *monacha.*

BOX 4.1 The Hubbs Family

Carl Leavitt Hubbs holds an honored place in the pantheon of American ichthyologists (for a detailed biography, see Shor et al., 1987). Born in 1894, he spent most of his childhood in the then-wilds of southern California, savoring the outdoors and honing his naturalist skills on animals ranging from marine mollusks to birds. His budding interests also included fish, and he eventually attended Stanford University, which was at that time a center for fish biology under the leadership of president David Starr Jordan (the monumental pioneer of ichthyology in North America). Hubbs graduated from Stanford in 1916; he took positions as museum curator of fish first at the University of Chicago (1917–1920) and subsequently at the University of Michigan (1920–1944). In 1944, he moved to the Scripps Institution of Oceanography, where he continued to study fish (and marine mammals) until his death in 1979, at the age of eighty-four. Hubbs produced more than 700 scientific publications in his lifetime, and in 1952 he was elected into the prestigious National Academy of Sciences. His work on fish and their environments has had important impacts on many fields, including taxonomy and systematics, archaeology, climatology, biogeography, natural history, ecology, evolution, conservation, and the history of science.

In 1918, Hubbs married Laura Clark, who became not only his lifelong companion but also a scientific collaborator on many of the family's ichthyologic expeditions and publications. The couple produced three children, including Frances, who married a famous ichthyologist (Robert Rush Miller), and Clark, who became a renowned ichthyologist in his own right. The extended Hubbs family has thus been at the center of American ichthyology for nearly a century. Although research on parthenogenetic fish is only a small part of Carl and Laura Hubbs' legacy, they were the first to identify and verify clonal reproduction in any vertebrate organism (*Poecilia formosa*), and their son Clark has a clonal fish (*Menidia clarkhubbsi*) named in his honor.

The *Po. monacha-lucida* complex is part of a broader assemblage of diploid hybridogens and triploid gynogens that arose in this same geographic region from additional crosses involving various sexual *Poeciliopsis* species (Schultz, 1961, 1967; Moore et al., 1970; Vrijenhoek et al., 1977, 1978). Some of the triploids are "trihybrid" gynogens that clonally propagate nuclear genomes from three sexual species (Vrijenhoek and Schultz, 1974). For example, *Po. monacha-lucida-viriosa*

BOX 4.2 Klaus Kallman and Tissue Transplantation Experiments

Before the advent of direct molecular assays of proteins and DNA, the most powerful method for distinguishing clones in various unisexual vertebrates involved tissue grafts. From research conducted primarily on mice, it was known that mammals possess refined histocompatibility systems, under the control of specific genes that dictate the outcome of experimental tissue transplants. Normally, skin cells or those of other tissues can be grafted successfully from one individual to another only if they do not possess any tissue antigens that are lacking in the cells of the host (Snell, 1957; Medawar, 1958; Billingham, 1959). In the late 1950s and early 1960s, Klaus Kallman began to publish reports showing that fish possess comparable histocompatibility loci (Kallman and Gordon, 1958; Kallman, 1960, 1961) and that these highly polymorphic genetic systems can help researchers detect and study vertebrate clonality (Kallman, 1962b). Even after molecular laboratory methods became available beginning in the mid-1960s, many researchers continued to use histocompatiblity bioassays as a powerful way to help distinguish clones in unisexual vertebrate taxa (e.g., Darnell et al., 1967; Moore, 1977; Angus and Schultz, 1979; Cuellar, 1984; Dawley et al., 2000).

The following synopsis of Kallman's scientific career is based on a commentary by James Atz and Steven Kazianis (2001). Born in Berlin in 1928, Kallman and his family survived the subsequent war and Russian occupation before immigrating to the United States, where he received his B.S. degree from Queens College in 1952. He then entered graduate school at New York University under the tutelage of Myron Gordon, who himself was an expert on fish genetics. In Gordon's lab, Kallman began to develop and refine tissue transplantation techniques (such as fin grafting) on small fish. He continued this laboratory effort even after Gordon's sudden death in 1959, eventually expanding his work into a model approach for studying a wide range of genetic and evolutionary phenomena related to vertebrate clonality (Kallman, 1975). Kallman worked on many other research challenges as well (e.g., sex-linkage in fish), and through more than 70 substantive publications has become recognized as one of the world's pioneering fish geneticists.

arose from successive hybridization events involving *Po. monacha, Po. lucida,* and *Po. viriosa* (Mateos and Vrijenhoek, 2005).

The name R. Jack Schultz is indelibly linked with clonal and hemiclonal fish, and in particular with unisexual members of the *Poeciliopsis* complex. His scientific legacy is large (box 4.3).

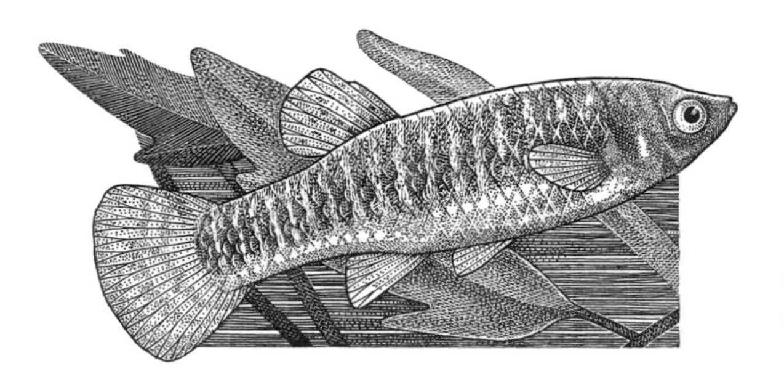

FIGURE 4.4 *Poeciliopsis monacha-lucida.*

Atherinidae (Silversides)

This globally distributed family includes more than 150 living species in about 30 genera. Most of the genera are marine, but some include freshwater representatives as well. Almost all of the species are sexual; only one clonal biotype is currently known.

Menidia clarkhubbsi

The Texas silverside (fig. 4.5) is a gynogenetic complex of all-female clones that arose as interspecific hybrids between males of *Me. beryllina* and females of an extinct or as-yet undetected species that genetically resembles extant *Me. peninsula* (Echelle and Echelle, 1997). This unisexual biotype is found along the Texas coast and eastward at sites on the northern Gulf of Mexico. It was discovered in the early 1980s (Echelle and Mosier, 1982; Echelle et al., 1983) and named in honor of a famous University of Texas ichthyologist (see box 4.1).

Cyprinidae (Minnows and Allies)

Members of this huge taxonomic assemblage are native to North America, Eurasia, and Africa. With more than 1,600 species in nearly 300 genera, this is the most species-rich family of fish (the closest competitor, Gobiidae, has about half as many species). Despite the abundance of cyprinid species and individuals, only a few clonal or hemiclonal biotypes are known within the family. These occur in three genera.

BOX 4.3 The Scientific Legacy of R. Jack Schultz

At the University of Michigan in the late 1950s, R. Jack Schultz (then a graduate student) and the famous ichthyologist Robert Rush Miller (see box 4.1) discovered all-female forms of *Poeciliopsis* fish in northwestern Mexico, and they announced their findings in the journal *Science* (Miller and Schultz, 1959). Gynogenetic fish in the related genus *Poecilia* (from northeastern Mexico) were already known by that time, but the newly discovered *Poeciliopsis* unisexuals showed genetic peculiarities that made them special (Schultz, 1961). Throughout his long research career, mostly at the University of Connecticut, Schultz studied unisexual *Poeciliopsis,* as well as some other clonal fish taxa, intensively. He was the first to successfully synthesize unisexual lineages in the laboratory (by crossing sexual species of *Poeciliopsis* experimentally; Schultz, 1973), he was responsible for initially working out many of the mechanistic details of hybridogenesis and gynogenesis that characterize this taxonomic assemblage (e.g., Schultz, 1967, 1969), and he was among the earliest writers to elaborate the potential ecological advantages and difficulties faced by unisexuals vis-à-vis their sexual counterparts (e.g., Schultz, 1971). Schultz retired about a decade ago. He has since been busy writing mystery novels, such as *The Fugitive at Greyledge* and *Charred Remains,* containing lots of murder and sex (perhaps to balance his earlier preoccupation with the paucity of sex in all-female fish).

Schultz's scientific legacy also includes his many students. Robert Vrijenhoek completed his Ph.D. under Schultz's direction in 1972. In a highly productive career at Rutgers University and later at the Monterey Bay Aquarium Research Institute, Vrijenhoek and his own students and collaborators have greatly expanded genetic and ecological knowledge about the *Poeciliopsis* complex, producing approximately 50 papers that have made these little fish a model system for studying unisexual vertebrates. Other prominent students of Schultz include Robert Dawley (who in 1989 coedited the first major book on unisexual vertebrates) and William Moore, Robb Angus, and Kate Goddard (all of whom have done seminal research on various clonal taxa).

Carassius auratus gibelio

By artificial means, such as irradiating and thereby disabling sperm used to fertilize eggs, scientists can induce gynogenesis in goldfish, *C. auratus* (Paschos et al., 2001; see chapter 7). Unisexual biotypes of *C. auratus* have arisen in the wild as well. The silver carp (fig. 4.6) is a gynogenetic form of triploid goldfish (*C.a.*

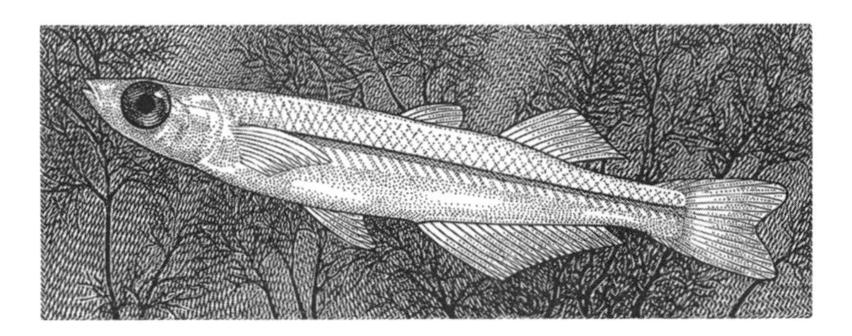

FIGURE 4.5 *Menidia clarkhubbsi,* the Texas silverside.

FIGURE 4.6 *Carassius auratus gibelio,* the silver carp.

gibelio) that occurs in nature as a series of distinct clones (Cherfas, 1966; Zhou et al., 2000a; Yang et al., 2001).

The silver carp also displays an interesting biological phenomenon with regard to sex. As many as 20% of all specimens in some natural populations of *C.a. gibelio* are males, and genetic evidence indicates that these individuals occasionally engage in recombination-generating sex with the otherwise gynogenetic females (Zhou et al., 2000b). How these males mechanistically arise is uncertain, but the recombinational genetic variation that they introduce to an otherwise clonal population could be ecologically important. Males have been reported in several other "asexual" vertebrate taxa as well (see beyond); the silver carp, however, may provide the first documented instance in which routine gonochorism (separate-sex reproduction) co-occurs with gynogenesis in an otherwise clonal biotype.

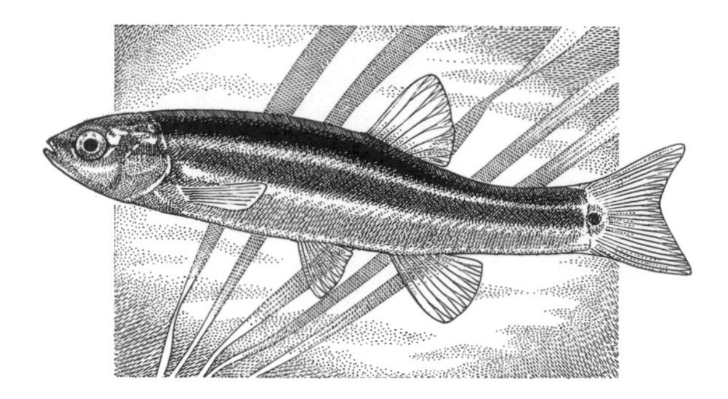

FIGURE 4.7 *Phoxinus eos-neogaeus,* a unisexual dace.

Phoxinus eos-neogaeus

The geographic range of these daces (small minnows; fig. 4.7) extends across most of the northern United States and southern Canada east of the Rocky Mountains. Carl Hubbs and D. E. S. Brown (1929) were the first to voice suspicion that apparent hybrids between *Ph. eos* and *Ph. neogaeus* might reproduce clonally, and decades later it was confirmed that some members of this assemblage are indeed gynogenetic and that a variety of unisexual biotypes exist (Dawley et al., 1987; Goddard and Schultz, 1993). Some of the female hybrids are diploid, some are triploids that carry two *eos* genomes and one *neogaeus* genome, some are triploids that carry two *neogaeus* genomes and one *eos* genome, and a few fish are even genetic mosaics with mixtures of diploid and triploid cells (Dawley and Goddard, 1988; Goddard et al., 1989).

The diploid hybrids comprise genuine gynogenetic lineages, but the triploids (which comprise approximately 50% of the unisexual *Phoxinus* specimens at some locations) do not. Instead, the latter arise de novo each generation when some diploid ova from gynogenetic females are fertilized by haploid sperm from a sexual parent species (*Ph. eos*). Among all gynogenetic vertebrates, this rate of sperm incorporation into eggs is unusually high. The triploids, as well as the mosaics, are typically fertile, but—unlike the diploid gynogens of *Ph. eos-neogaeus* or the triploid gynogens in some other unisexual vertebrates—they are thought not to reproduce clonally.

Leuciscus (or *Rutilus* or *Tropidophoxinellus*) *alburnoides*

These small minnows (fig. 4.8) are successful and abundant in rivers of the Iberian Peninsula in Europe, their native range. The "species" encompasses a

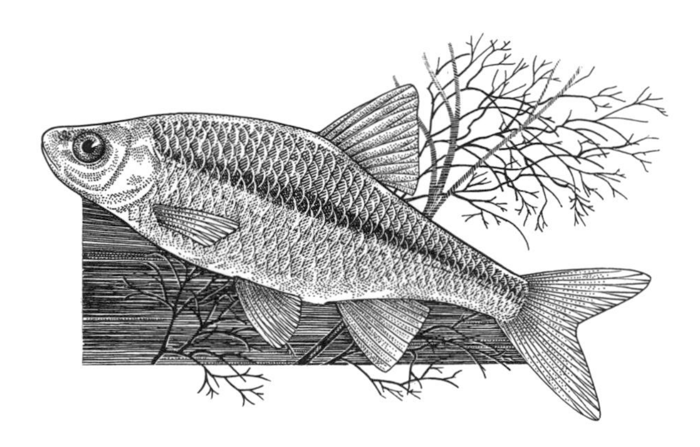

FIGURE 4.8 *Leuciscus alburnoides,* an Iberian minnow.

bewildering variety of genetic and reproductive biotypes that arose via multiple hybridization events involving the bisexual species *Le. pyrenaicus* (in southern rivers) or *Le. carolitertii* (northern drainages) and at least one other yet-unspecified bisexual species (Collares-Pereira, 1989; Alves et al., 1997, 2001). Triploid hybridogenetic females predominate at most locations; also present in the taxonomic assemblage are diploid hybridogenetic females, as well as diploid, triploid, and tetraploid males that can be fertile and reproductive (Carmona et al., 1997; Alves et al., 2002). Adding to this complexity is a diversity of genetic mechanisms besides traditional hybridogenesis (table 4.1), including the production of haploid and polyploid gametes by various classes of males and females, and frequent syngamy (successful unions between sperm and egg). These reproductive modes make *Le. alburnoides* difficult to pigeonhole in terms of the categories traditionally used to describe "unisexual" vertebrates. They also imply that genetic exchanges and evolutionary shifts between various genetic forms of *Le. alburnoides* are frequent, if not routine, and that the mostly female hybrid complex in effect has many of the potential advantages traditionally associated with sexual reproduction (Alves et al., 1998).

Cobitidae (Loach Fish)

About 150 extant species of these freshwater fish inhabit Eurasia, with species diversity being highest in southern Asia. These fish are typically bottom-dwellers with downward-facing mouths and wormlike or fusiform (spindle-shaped) bodies. In one genus, *Cobitis*, multiple hybridization events involving several bisexual

TABLE 4.1 A remarkable diversity of genetic features and reproductive modes characterizes various biotypes of the *Leuciscus alburnoides* complex of Iberian minnows[a]

Ploidy level	*Sex*	*Genome constitution*[b]	*Reproductive mode or feature*	*Ploidy of gametes*
2N	female	*c/u*	hybridogenesis	N
2N	female	*p/u*	production of unreduced eggs with high levels (nearly 100%) of syngamy	2N
2N	male	*p/u*	production of fertile, unreduced sperm	2N
2N	male	*u/u*	normal meiosis	N
3N	female	*c/u/u*	hybridogenesis	2N
3N	female	*p/u/u*	hybridogenesis[c]	N, 2N
3N	male	*p/u/u* or *p/p/u*	fertility unknown	3N
4N	male	*p/p/u/u*	normal meiosis	2N

[a]Adapted from Alves and colleagues (2001).
[b]*p* = a haploid genome from *Le. pyrenaicus; c* = a haploid genome from *Le. carolitertii; u* = a haploid genome from an unknown ancestor.
[c]Albeit by a different cellular mechanism than for the other hybridogenetic forms listed.

species—including *taenia, elongatoides, tanaitica, sinensis,* and *longicarpus*—have given rise to a series of mostly diploid and triploid clonal biotypes that usually reproduce by gynogenesis (Vasil'ev et al., 1989; Kim and Lee, 1990, 2000; Janko et al., 2003, 2007), although a report of hybridogenesis exists as well (Bohlen et al., 2002). Clonal reproduction has also been confirmed for some specimens in another cobitid genus, *Misgurnus* (Morishima et al., 2002, 2004, 2007).

Cobitis elongatoides-taenia

Like other members of this gynogenetic complex of spined loaches, these fish (fig. 4.9) inhabit river drainages of central Europe that drain into the North Sea, Baltic Sea, or Black Sea. This particular taxon is polyphyletic, having arisen via hybrid crosses in both directions between its two bisexual progenitor species. Although triploids predominate in most populations, diploid and tetraploid hybrids are also known (Vasil'ev et al., 1989; Janko et al., 2007).

Misgurnus anguillicaudatus

This Japanese loach (fig. 4.10) normally reproduces as a bisexual diploid, but genetic analyses of wild populations have revealed instances of otherwise cryptic

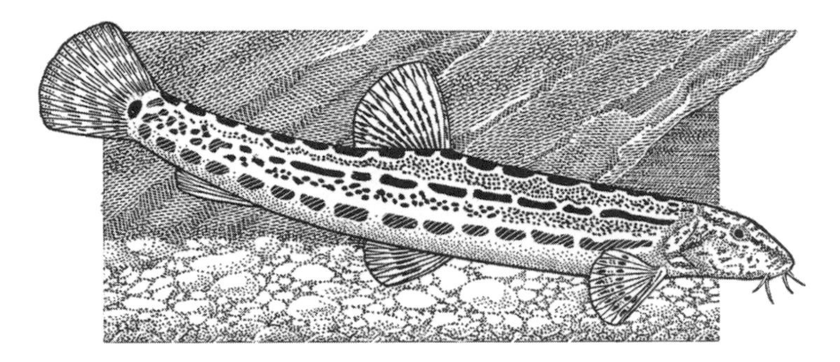

FIGURE 4.9 *Cobitis elongatoides-taenia*, a spined loach.

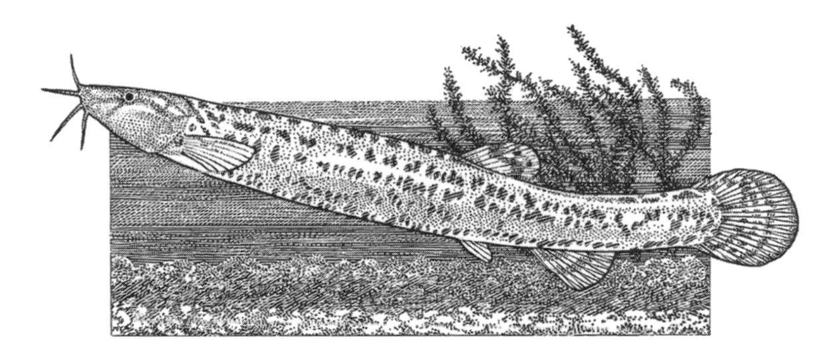

FIGURE 4.10 *Misgurnus anguillicaudatus*, a Japanese loach.

clonal reproduction as well. This happens when females occasionally produce unreduced diploid eggs that undergo successful development into new individuals, without genetic input from sperm. Furthermore, in some other cases the sperm cell actually does fertilize the diploid egg, resulting in the natural occurrence of triploid specimens.

Cyprinodontidae (Killifish)

The family Cyprinodontidae contains more than 300 species, distributed mostly in fresh and brackish waters throughout the Americas, Africa, and Eurasia. Killifish are egg layers, and with few exceptions they reproduce by standard bisexual means. One of the few exceptions involves an all-female clonal biotype within the otherwise sexual species *Fundulus heteroclitus* (fig. 4.11). This biotype, which arose via hybridization between standard sexual forms of *F. heteroclitus*

FIGURE 4.11 *Fundulus heteroclitus,* the mummichog.

and *F. diaphanus,* is known from two sites in Nova Scotia, Canada, and is probably gynogenetic (Dawley, 1992).

Ambystomatidae (Mole Salamanders)

The common name, "mole salamanders," for this family comes from its members' habit of burrowing in forest humus or hiding under logs or rocks for most of their lives, the main exception being when these nocturnal amphibians emerge after a seasonal rain to congregate and mate in temporary pools and ponds. Females lay eggs that become attached to underwater sticks or other debris. The aquatic larvae are carnivorous (like the adults), with wide heads, large tailfins, and long feathery gills. Adult mole salamanders have relatively stocky bodies and limbs, short broad heads, small eyes, smooth skin, and flattened tails. The family is strictly North American, and its most prominent genus (*Ambystoma*) contains about 30 bisexual species distributed from the Central Valley of Mexico to Alaska in the west and Labrador in the east.

Also in the genus *Ambystoma* are diverse unisexual biotypes, the first of which were discovered by Wesley Clanton in 1934 and further characterized by S. A. Minton in 1954 (reviewed in Bogart and Klemens, 1997). These mostly-female "unisexuals"—in which males sometimes occur in very low frequency, too—are abundant in the Great Lakes region of North America (Licht, 1989; Bogart and Klemens, 1997).

Karyotypes of these unisexual salamanders range from diploid through pentaploid, with the nuclear genomes referable in each case to a particular pair, triplet, or quadruplet among four bisexual species: *Am. laterale, Am. jeffersonianum, Am. texanum,* and *Am. tigrinum* (Uzzell, 1964; Bogart, 1989; Lowcock, 1989; Kraus et al., 1991). For example, the triploid unisexual "LLT" (Lowcock et al., 1987) carries two copies of the *laterale* nuclear genome plus one copy from *texanum;* the tetraploid unisexual LTJTi carries one copy each of the *laterale, texanum, jeffersonianum,* and *tigrinum* haploid genomes. Each of the more than 20 different unisexual biotypes identified to date has at least one copy of the *laterale* nuclear genome, but the sexual species that contributed the additional nuclear genome(s) varies from case to case. A few of the unisexuals also have numbers of major chromosomes that are not exact multiples of the typical haploid set ($N = 14$) for *Ambystoma,* and these individuals are reportedly the first such aneuploid unisexuals known to science (Bi et al., 2007b).

The reproductive systems of various *Ambystoma* unisexuals have been referred to as parthenogenetic (Uzzell, 1969; Downs, 1978), gynogenetic (Macgregor and Uzzell, 1964, Elinson et al., 1992; Hedges et al., 1992), and hybridogenetic (Bogart et al., 1989; Hedges et al., 1992), but in fact the genetics are extremely complex and not readily pigeonholed, overall, into any one of those traditional categories. James Bogart and colleagues (2007) introduced the term "kleptogenetic" for these salamanders, for reasons that will be discussed later.

Ambystoma laterale-texanum

This unisexual biotype (fig. 4.12), known only from islands and mainland sites at the western end of Lake Erie (Downs, 1978; Kraus, 1985, 1989), carries haploid nuclear genomes of the sexual species *Am. laterale* and *Am. texanum.*

FIGURE 4.12 *Ambystoma laterale-texanum,* a unisexual mole salamander.

Ranidae ("True Frogs")

Members of Ranidae occur on all continents except Antarctica. These amphibians are generally long-legged and smooth-skinned, and they have toes that are joined by webs but forelimb fingers that are free. Males have vocal pouches under the throat or at its sides. The thumbs of breeding males are enlarged, for grasping females during amplexus (mating).

The largest genus in the family is *Rana,* with more than 250 bisexual species. One of these—*R. ridibunda*—has produced hybridogenetic derivatives by hybridizing with various other *Rana* species.

Rana esculenta

This complex of hybridogenetic water frogs (fig. 4.13), with a range extending across much of central Europe, arose from crosses between the bisexual species *R. ridibunda* and *R. lessonae* (Uzzell and Berger, 1975; Graf and Polls Pelaz, 1989). Some specimens are diploid, but others are triploid and carry either two copies of the *lessonae* nuclear genome and one of *ridibunda,* or vice versa. One striking feature that distinguishes these hybridogenetic frogs from most other clonal or hemiclonal vertebrates is a high incidence of males; indeed, some populations consist of males only. These individuals generally have low reproductive fitness, but they can mate with females of a sexual species and hemiclonally transmit (via sperm) one genome to their progeny (Graf and Polls Pelaz, 1989). In other populations, only female *esculenta* reproduce, hemiclonally transmitting one

FIGURE 4.13 *Rana esculenta,* a hybridogenetic water frog.

genome to their offspring (Uzzell et al., 1980). In some cases, hybridogenetic males and females also mate with one another to produce all-hybrid populations (see Dawley, 1989). Additional variations on these themes exist (Graf et al., 1977; Hotz and Uzzell, 1983; Hotz et al., 1985; Ragghianti et al., 2007)—such as which parental genome is excluded and which is hemiclonally retained (Uzzell et al., 1977)—thus making *R. esculenta* an extremely complicated genetic system overall.

Cellular and Genetic Mechanisms

Gynogenesis

At least two distinct cellular mechanisms underlie gynogenesis in clonal vertebrates that have been examined in cytological detail. The first is premeiotic endomitosis; this was discussed earlier (see fig. 3.17) for parthenogenetic *Aspidoscelis* lizards but also applies, for example, to triploid *Poeciliopsis* fish that reproduce gynogenetically (Cimino, 1972a), as well as to unisexual *Ambystoma* salamanders (Macgregor and Uzzell, 1964) and *Misgurnus* loaches (Itono et al., 2006). The second is apomixis, an oogenic process that is entirely mitotic, that is, without hints or vestiges of meiosis-type events. Under apomixis (which in essence is conventional mitosis as diagrammed earlier in fig. 1.3), chromosomes in a precursor egg cell duplicate; however, the homologues do not synapse or engage in crossing over before distributing equally to ova whose genomes are thus identical to those of the parent's somatic cells. Each egg then develops directly into a new individual who is clonally identical to her mother and sisters. Apomixis is common in many organisms, including various plants, but this mechanism also characterizes at least one group of clonal vertebrates—the gynogen *P. formosa* and its triploid unisexual derivatives (Rasch et al., 1982; Monaco et al., 1984).

A recent genetic analysis suggests that a form of automixis can also occur in *Poecilia* fish. Based on inheritance patterns observed in strains of *formosa*-like fish experimentally synthesized in the laboratory (by crossing *P. mexicana* with *P. latipinna*), Kathrin Lampert and colleagues (2007) deduced that meiotic cellular divisions followed by the fusion of meiotic products take place routinely in the hybrid females, the net result being diploid eggs with various recombinations of maternally derived genes. Whether automixis ever occurs in *P. formosa* or other unisexual taxa in nature remains to be determined. The phenomenon is unlikely, though, to be the standard reproductive mode in most self-sustaining unisexual biotypes (at least those that are diploid), because individuals in such strains are tyically highly heterozygous (whereas heterozygosity should be lost quickly under automixis, which in effect is an intense form of inbreeding). The possibility nonetheless remains that *P. formosa* or some other gynogens in nature might have arisen by automixis but subsequently been maintained by apomictic gynogenesis (Lampert et al., 2007).

Hybridogenesis

In classic hybridogenesis, the paternal chromosome set is excluded from each functional egg during oogenesis. So the maternal genome is the clonal component of this hemiclonal system; fertilization by a sperm cell then reestablishes a nonreduced genotypic condition in each resulting offspring. Oddly, each hybridogenetic female expresses (functionally utilizes) the paternal genome that she inherited from her father, yet she fails to transmit that paternal genome to her daughters.

In hybridogenetic taxa that have been examined in cytological detail, at least two cellular mechanisms have been found to underlie the exclusion of the paternal genome. One of these mechanics is illustrated by *Po. monacha-lucida*. During a mitotic division that precedes meiosis, spindle fibers from one pole of a diploid cell attach to *monacha* chromosomes, but not to *lucida* chromosomes that thus languish in the cytoplasm and eventually are lost (Cimino, 1972b). Next, the *monacha* chromosomes enter a modified meiosis that is necessarily atypical because *lucida* chromosomes are unavailable for crossing over and segregation. The final cell division in this process thereby yields haploid eggs containing only the chromosomes of *monacha* origin.

A second form of genome exclusion is illustrated by diploid *R. esculenta*. Again, one set of parental chromosomes (usually from *R. lessonae* in this case) is lost premeiotically, but then the remaining chromosomes (from *R. ridibunda*) are duplicated by an endomitotic event such that they enter the first stage of meiosis as pairs of identical partners. Although these pairs engage in crossing over, genetic recombination in effect is suppressed because the synapsed chromosomes are clonal copies of one another. The second stage of meiosis then produces haploid gametes that carry only the *ridibunda* genome. In triploid specimens of *R. esculenta,* one or two entire genomes are again excluded during gametogenesis, yielding gametes that are either haploid or diploid. The mechanics of this process are not fully understood.

Interestingly, the hybridogenetic mechanism in diploid *R. esculenta* seems not to be as stable as that in *Po. monacha-lucida*. Occasional diploid gametes are produced, and so, too, are some recombinant haploid gametes that carry bits of paternally derived nuclear DNA (Uzzell et al., 1977). Similar types of genetic complexities, including sporadic genetic recombination between maternal and paternal chromosomes, probably apply to various other hybridogenetic vertebrate biotypes as well (Alves et al., 2001; Mateos and Vrijenhoek, 2002). Accordingly, at least some hybridogenetic lineages should be considered quasi-clonal (or quasi-sexual) not only because of the standard syngamy that takes place between ova and sperm in each generation, but also because some paternal DNA may leak into the otherwise hemiclonal system via sporadic recombination events during gametogenesis.

Kleptogenesis

In the unisexual *Ambystoma* salamanders, females sometimes reproduce gynogenetically. Genetic markers (allozymes, microsatellites, and chromosome segments) indicate, though, that the unisexual biotypes also routinely incorporate nuclear DNA from locally sympatric sexual species into their diploid or polyploid nuclei. Furthermore, they suggest that these male-derived nuclear genomes are neither kept intact nor eliminated at the ensuing meiotic event (Bi and Bogart, 2006; Bogart et al., 2007). Thus the genetic system of the unisexual biotypes is not strictly gynogenetic (or hybridogenetic or parthenogenetic), but rather is a unique reproductive mode for which a new term may be desirable. Alain Dubois and Rainer Günther (1982) proposed "klepton" (from kleptomania, a compulsion for theft) for such unisexual taxa, and following on that idea, James Bogart and colleagues (2007) coined the word "kleptogenetic" to refer to *Ambystoma* biotypes that routinely "steal" genetic material from sympatric sexual species.

The kleptogenetic system of *Ambystoma* is not fully understood, but the unisexuals apparently have a flexible genetic system that includes the capacity to generate and perpetuate (across the generations) various intergenomic exchanges with sympatric sexual relatives (Bi and Bogart, 2006; Bi et al., 2007a). Kleptogenesis in the mole salamanders may be merely an extreme example of the more petty genetic larcenies by which some other unisexual lineages—such as in *Rana* frogs or *Leuciscus* fish—occasionally incorporate DNA from related sexual species. Any such "kleptogen" is therefore not strictly clonal or hemiclonal, instead probably gaining some of the benefits of sexual recombination. This fact that should be borne in mind when interpreting the local abundance and ecological success that some of these biotypes, including *Ambystoma* (Bogart and Klemens, 1997), enjoy.

Evolution and Phylogeny

Hybrid Origins

All known gynogenetic and hybridogenetic lineages in vertebrates apparently originated via hybridization events between bisexual species. The evidence is normally unambiguous. For example, gynogenetic strains of the minnow *Ph. eos-neogaeus* are heterozygous at all surveyed allozyme loci for which their suspected parent taxa, the sexual species *Ph. eos* and *Ph. neogaeus,* display fixed genetic differences (Goddard and Dawley, 1990). Likewise, diploid forms of *P. formosa*—a gynogenetic live-bearing fish—display nearly fixed heterozygosity at numerous protein-coding loci that distinguish or are polymorphic in its sexual parent species (Abramoff et al., 1968; Balsano et al., 1972; Turner, 1982).

So each somatic cell of a gynogenetic or hybridogenetic individual houses at least two distinct nuclear genomes and is therefore highly heterozygous. Abnormal functional interactions between these amalgamated heterospecific genomes probably account, in general, for why clonal or hemiclonal taxa have such peculiar gametogenic (gamete-forming) cellular mechanisms. In a prototypical gynogenetic lineage, the full ensemble of two or more heterospecific genomes is transmitted clonally across animal generations; in a prototypical hybridogenetic lineage, one homospecific genome is passed more or less clonally to an ovum, which is then fertilized by sperm from a second species to reestablish a heterospecific genomic combination in offspring.

Parent Sexual Species and Direction of Cross

Scientists have used genetic markers from nuclear and mitochondrial genomes to help disclose the bisexual species and directions of hybridization that produced various gynogenetic and hybridogenetic biotypes (table 4.2). The successful crosses were in most cases apparently unidirectional with respect to sex, rather than reciprocal. For example, all surveyed unisexual biotypes in *Poeciliopsis* carry *monacha*-type mtDNA and thus had *Po. monacha* as the original maternal parent. There are, nonetheless, some exceptions to nonreciprocal matings: in both the *Co. elongatoides-taenia* and the *R. esculenta* complexes, hybridization events that contributed genes to the clonal or hemiclonal biotypes took place in both directions with respect to sex (table 4.2).

The *Rana* case merits elaboration. *Rana esculenta* is nearly unique among "unisexual" vertebrates in consisting of males as well as females in high frequency. From behavioral considerations, the initial hybrid events that produced *R. esculenta* probably involved male *R. lessonae* × female *R. ridibunda,* but once the hybridogens were formed, occasional matings of male *R. esculenta* × female *R. lessonae* may secondarily have introduced *R. lessonae* mtDNA into the *R. esculenta* biotype (Spolsky and Uzzell, 1986). Moreover, females belonging to such *R. esculenta* lineages appear to have served as a natural genetic bridge for the transfer of *R. lessonae* mtDNA into particular *R. ridibunda* populations via matings with *R. ridibunda* males (Spolsky and Uzzell, 1984).

Another complex scenario was hypothesized for the maternal ancestry of the triploid salamander *Am. 2-laterale-jeffersonianum,* which like the other unisexual *Ambystoma* oddly carries mtDNA from *Am. texanum.* Fred Kraus and Michael Miyamoto (1990) have proffered an explanation in which an original *Am. laterale-texanum* hybrid female produced an ovum with *Am. laterale* nuclear DNA primarily but with a female-determining sex chromosome (W) and the mtDNA

TABLE 4.2 Examples of gynogenetic and hybridogenetic biotypes for which the parental species and direction of the original hybrid crosses have been identified using a combination of nuclear and mitochondrial markers

		Sexual parental species		
Clonal or hemiclonal biotype	*Ploidy level*	*Males*	*Females*	*References*
Poecilia fish				
formosa	2N	*latipinna*	*mexicana*	Avise et al. (1991)
Poeciliopsis fish				
monacha-lucida	2N	*lucida*	*monacha*	Avise and Vrijenhoek (1987); Quattro et al. (1991)
monacha-occidentalis	2N	*occidentalis*	*monacha*	Quattro et al. (1992a)
2 monacha-lucida	3N	*lucida*	*monacha* (2)	Quattro et al. (1992b)
monacha-2 lucida	3N	*lucida* (2)	*monacha*	Quattro et al. (1992b)
Phoxinus fish				
eos-neogaeus	2N, 3N	*eos*	*neogaeus*	Goddard et al. (1989)
Menidia fish				
clarkhubbsi	2N	*beryllina*	*peninsula*	Echelle et al. (1989a, 1989b)
Leuciscus fish				
alburnoides complex	2N–4N	uncertain	*pyrenaicus*	Alves et al. (1997)
Cobitis fish				
elongatoides-tanaitica	2N, 3N	*tanaitica*	*elongatoides*	Janko et al. (2003)
elongatoides-taenia	2N, 3N	*taenia, elongatoides*	*elongatoides, taenia*	Janko et al. (2003)
Ambystoma salamanders				
unisexual complex	2N–5N	*laterale, texanum, tigrinum, jeffersonianum*	*barbouri*[a]	Bogart (2003); Bogart et al. (2007)
Rana frogs				
esculenta	2N	*lessonae, ridibunda*	*ridibunda, lessonae*	Spolsky and Uzzell (1986)

[a]Formerly considered conspecific with *texanum* (see Kraus and Petranka, 1989).

of *Am. texanum.* When fertilized by a male *Am. laterale,* female progeny with two *Am. laterale* nuclear genomes and the mtDNA of *Am. texanum* resulted. Subsequent hybridization with male *Am. jeffersonianum* then produced the *Am. 2-laterale-jeffersonianum* biotypes now carrying *texanum*-like mtDNA. This origination scenario remains speculative, but its mere feasibility suggests that odd or idiosyncratic hybridization pathways probably underlie the geneses of some unisexual taxa.

For all known vertebrates that display gynogenetic and hybridogenetic reproduction, the respective sexual ancestors are invariably deemed congeneric. Seldom, though, have they proved to be sister taxa. For example, *Po. monacha* and *Po. lucida,* which via hybridization generate unisexual lineages, are each more closely related to various other *Poeciliopsis* species than they are to one another (fig. 4.14).

Such phylogenetic findings are consistent with the balance hypothesis (see chapter 3), which posits that the genomes of hybridizing species that yield clonal derivatives are divergent enough to disrupt gametogenic mechanisms in progeny, yet not so divergent as to seriously compromise hybrid viability or fertility. The phylogenetic findings are also consistent, however, with the phylogenetic constraint hypothesis, which posits that genetic peculiarities predispose particular bisexual species to produce clonal lineages via hybridization. Regardless of which hypothesis is correct, all gynogenetic and hybridogenetic (as well as parthenogenetic) vertebrates currently known to science clearly have arisen by interspecific crosses involving only a small subset of all bisexual congeners that are potentially capable of hybridizing.

Triploid Mechanics

Some sperm-dependent unisexual lineages are diploid and others are polyploid (usually triploid). Because gynogens and hybridogens must mate to reproduce, their eggs are routinely exposed to sperm, and this probably makes the production of triploids even more likely than in parthenogens. In gynogenetic *P. formosa,* for instance, syngamy (fertilization by sperm) occurs in about 1% of the diploid eggs (Schultz and Kallman, 1968; Rasch et al., 1970); in *Poeciliopsis* fish, females occasionally produce diploid (rather than haploid) eggs that, when fertilized, may yield gynogenetic triploid strains (Cimino, 1972a).

Triploid unisexual biotypes generally arise as depicted earlier in figure 3.18, but careful genetic inspection has sometimes revealed further details about the cellular mechanics. One issue is whether the unreduced diploid egg that became fertilized by a haploid sperm was produced by a nonhybrid diploid female (the spontaneous-origin model) or, alternatively, by a hybrid diploid female (the primary-hybrid model). As discussed previously in the context of parthenogenesis (see fig. 3.21), these competing hypotheses can be tested by cytonuclear appraisal.

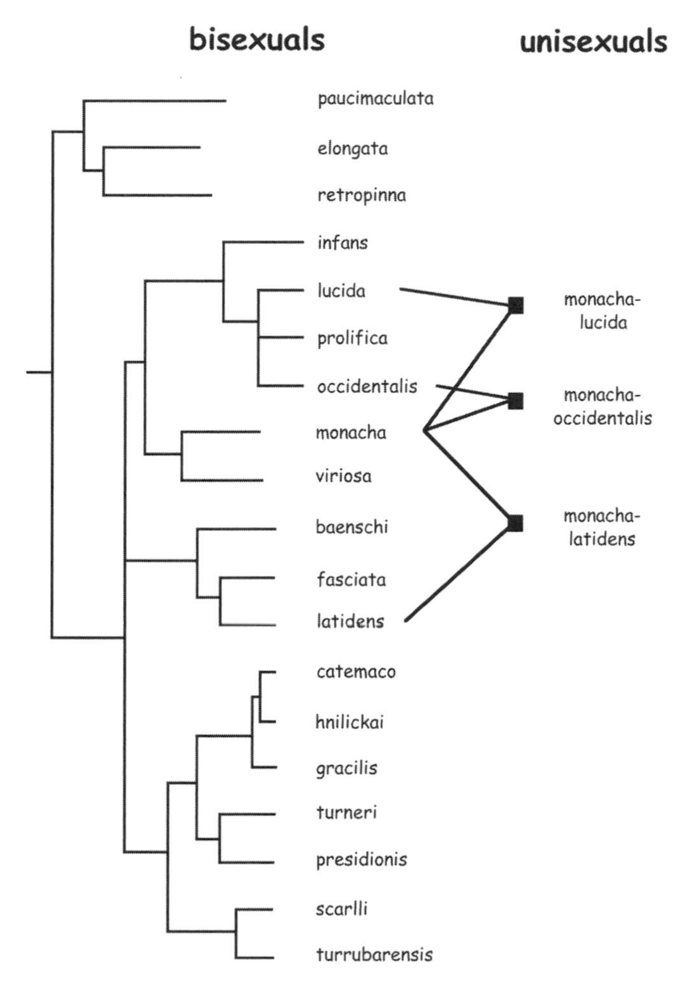

FIGURE 4.14 Phylogenetic estimate for nearly all of the 21 named species of bisexual *Poeciliopsis,* based on extensive sequence comparisons of mtDNA (Mateos et al., 2002). Also shown are primary crosses that have been involved in generating various unisexual biotypes.

For a gynogenetic triploid, the spontaneous-origin scenario predicts that the paired homospecific nuclear genomes derive from the sexual species that was the maternal parent in the original hybridization, and thus should be coupled with mtDNA from that same parent species. The most likely version of the primary-hybrid scenario, on the other hand, predicts that paired homospecific genomes in a gynogenetic line trace back to the sexual species that was the male parent in the initial cross, and thus should not be paired with mtDNA from that same parent species.

Joseph Quattro and colleagues (1992b) examined a triploid *Poeciliopsis* gynogen with these hypotheses in mind, and found that *Po. monacha-2 lucida* possesses the mtDNA of *Po. monacha* but two nuclear genomes from *Po. lucida,* thus refuting the spontaneous-origin model for these unisexual fish. Together with findings for triploid parthenogenetic lizards (genus *Aspidoscelis;* see chapter 3), these results support the long-standing view that interspecific hybridization is often the initial trigger for the atypical gametogenic mechanisms that can lead not only to diploid unisexual biotypes but to triploid unisexual lineages, too (Schultz, 1969).

Assuming that triploid unisexuals usually arise via the primary-hybrid route, two further cytogenetic pathways to triploidy can be distinguished (fig. 4.15). Under the genome-addition scenario (Schultz, 1969), interspecific F_1 hybrids produce unreduced ova (AB) that, when fertilized by sperm from one of the sexual ancestors, yield allotriploid biotypes (AAB or ABB). Under the genome-duplication scenario (Cimino, 1972a), suppression of an equational division in an F_1 hybrid produces unreduced ova (AA or BB) that, when fertilized by sperm from species A or B, would also yield AAB or ABB progeny. (In principle, this latter process could also yield autotriploid AAA or BBB progeny, but no such self-sustaining unisexual lineages are known in vertebrates.) A key distinction between these two pathways involves predicted levels of heterozygosity in the two homospecific nuclear genomes. In the genome-duplication model, heterozygosity should be extremely low—the only variation being derived from postformational mutation—whereas heterozygosity in the genome-addition model should be about the same as that in the sexual species.

In molecular assays of allozyme loci, Quattro and colleagues (1992b) found that the two paired nuclear genomes of triploid *Poeciliopsis* gynogens show normal levels of heterozygosity, effectively falsifying the genome-duplication hypothesis for these fish. Genome addition is probably also responsible for the origin of other gynogenetic triploid fish, including at least some populations of *P. formosa* (Turner et al., 1980), *Me. clarkhubbsi* (Echelle et al., 1988), and *Ph. eos-neogaeus* (Goddard et al., 1989).

Tetraploid specimens are generally less common in unisexual complexes, but they do occur in several groups such as *Ambystoma* salamanders and *Cobitis* and *Leuciscus* fish. For the *Le. alburnoides* unisexual assemblage, tetraploid individuals reportedly arise when diploid sperm produced by diploid hybrid males occasionally fertilize diploid eggs (Alves et al., 1999), or when occasional triploid eggs produced by triploid females are fertilized by haploid sperm (Alves et al., 2004). Elevated temperature has been shown to increase the frequency of tetraploids in unisexual *Ambystoma* (Bogart et al., 1989).

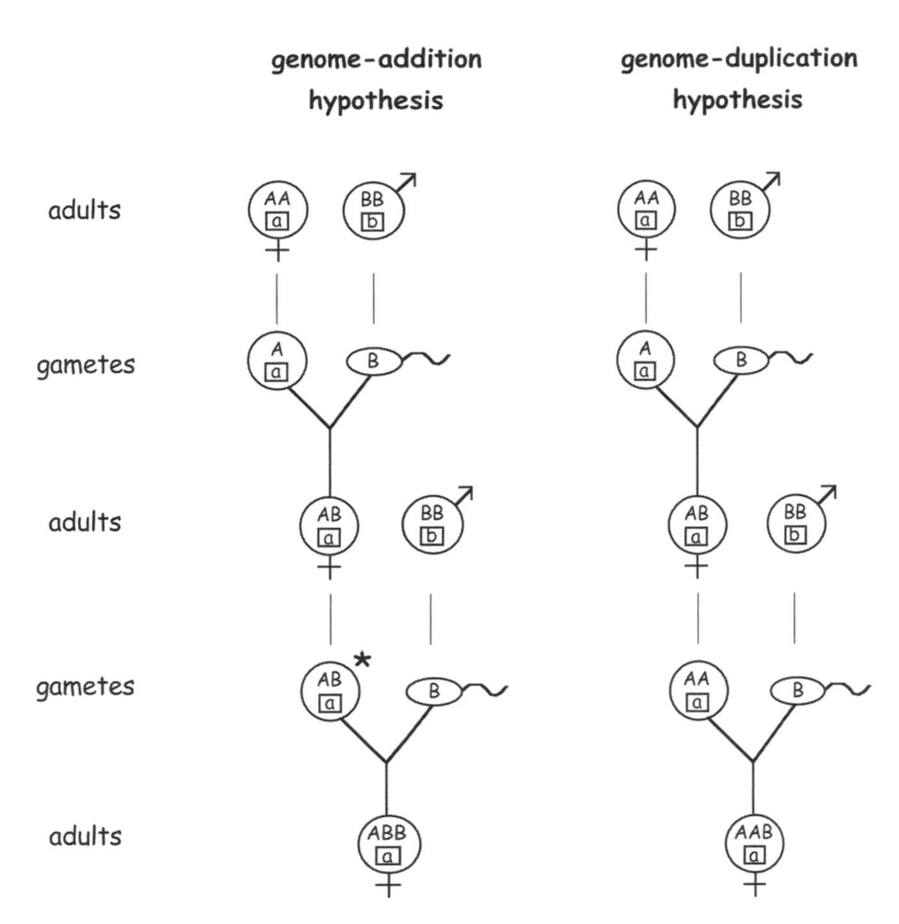

FIGURE 4.15 Schematic diagram of competing mechanisms for the origin of a triploid unisexual lineage, assuming correctness of the primary-hybrid model (see fig. 3.21). Each uppercase letter represents one nuclear genome (A or B) from the respective parent species, and the lowercase letters in boxes similarly refer to the maternally inherited mtDNA genome. Ovals and circles indicate sperm and eggs, respectively, the latter being unreduced where indicated by stars.

Number of Hybridization Events

Notwithstanding several complications discussed earlier in box 3.4, genetic data can be used to estimate the approximate number of hybridization events that have produced each gynogenetic or hybridogenetic "species." For example, multilocus allozyme surveys revealed that the taxon *Me. clarkhubbsi* consists of at least half a dozen gynogenetic clones that probably register separate hybrid geneses rather than postformational mutations in a monohybrid line (Echelle, Dowling, et al., 1989; Echelle, Echelle, et al., 1989; Echelle and Echelle, 1997). Qualitatively similar conclusions about multiple hybrid origins have been reached from molecular genetic surveys of the *Cobitis* complex of gynogenetic spined loaches (Janko et al., 2005), the *Ph. eos-neogaeus* assemblage of gynogenetic dace (Binet and Angers, 2005; Angers and Schlosser, 2007), and the *Le. alburnoides* complex of hybridogenetic minnows (Alves et al., 2001). Conversely, based on the limited genetic diversity observed among individuals of gynogenetic *P. formosa,* the possibility of a single hybridization origin for the diploid forms of this unisexual biotype cannot be eliminated (Schartl et al., 1995b).

For unisexual salamanders in the genus *Ambystoma,* allozyme data originally were interpreted as reflecting multiple origins through recurring hybridization and backcrossing (Bogart et al., 1985, 1987; Bogart and Licht, 1986; Lowcock and Bogart, 1989). However, the recent demonstration of ongoing genetic exchange with sympatric bisexual species complicates interpretations about the exact number (and even the true meaning) of the original evolutionary geneses in these cases. Furthermore, at least some restrictions on the initial genesis process are implied by the uniform mtDNA origin (see table 4.2) of all of the *Ambystoma* unisexuals (Bogart, 2003; Bogart et al., 2007).

Poeciliopsis fish provide well-documented examples of how recurring hybridization can produce multiple unisexual lineages. In 1973, Jack Schultz announced an amazing accomplishment: he had generated hybridogenetic fish in the laboratory. By forcing two bisexual species (*Po. monacha* and *Po. lucida*) to cross, Schultz had spontaneously synthesized several all-female hybridogenetic lineages that seemed entirely comparable to those already known from wild populations of *Po. monacha-lucida* in Mexico. Moreover, the hybridogens generated in Schultz's laboratory emerged exclusively from crosses of *Po. monacha* females × *Po. lucida* males. This same direction of hybridization would later be shown (by cytonuclear DNA evidence) to have generated all of the natural strains of *Po. monacha-lucida* in Mexico (Avise and Vrijenhoek, 1987; Quattro et al., 1991).

In Schultz's original laboratory experiments, five of 67 matings (7%) of *Po. monacha* with *Po. lucida* were successful in the sense of yielding viable and fertile hybridogenetic offspring. Thus, although the genetic hurdles to the spontaneous formation of unisexual biotypes may be formidable, they seem to be far from insurmountable in these fish. Molecular genetic surveys and other evidence revealed that natural hybridogenetic lineages of *Poeciliopsis* in Mexico likewise originated from multiple hybridization events (Quattro et al., 1991). Such observations give added vindication for the frozen-niche-variation model (Vrijenhoek, 1979, 1984a, 1984b), which posits that extensive interclonal genetic variation due to recurrent hybridization is available for scrutiny by natural selection in these fish. The model will be elaborated later in this chapter.

Genealogical Histories

Mitochondrial genotypes, even in the absence of data from nuclear DNA, sometimes provide clear evidence that a unisexual taxon traces back to more than one hybrid genesis. For example, hybridogenetic lineages of *Po. monacha-lucida* are interspersed across the matrilineal genealogy of their bisexual mother species, *Po. monacha* (fig. 4.16), such that these unisexuals are clearly polyphyletic, collectively, in maternal ancestry. A qualitatively similar outcome (polyphyly) has been reported for about 50% of the gynogenetic and hybridogenetic taxa genetically surveyed for mtDNA (Avise et al., 1992), including *Me. clarkhubbsi* (Echelle et al., 1989a, 1989b), *Ph. eos-neogaeus* (Goddard et al., 1989), and *R. esculenta* (Spolsky and Uzzell, 1986).

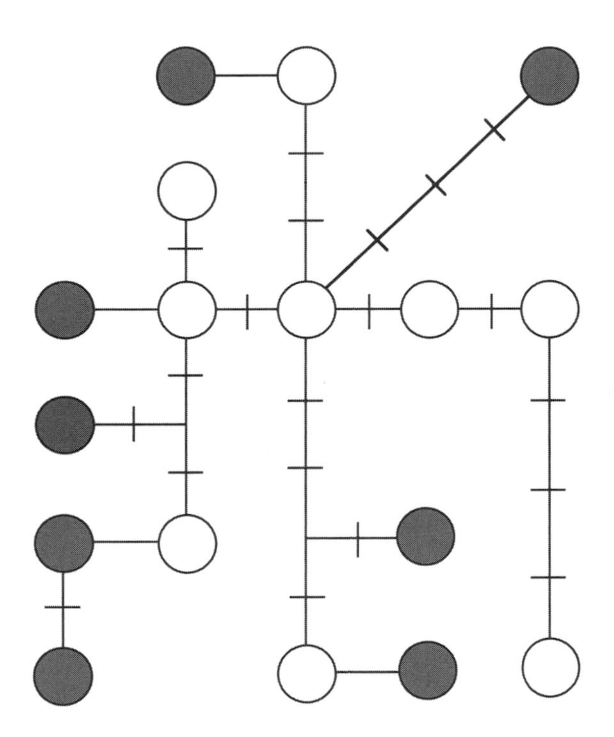

FIGURE 4.16 Phylogenetic network for the mtDNA genotypes observed in natural populations of sexual *Poeciliopsis monacha* (open circles) and unisexual *Po. monacha-lucida* (shaded circles) (after Quattro et al., 1991). Slashes represent the number of deduced mutations along each branch of the evolutionary network.

In several other cases, however, a gynogenetic or hybridogenetic biotype has proved to be just one small genealogical subset nested within the broader matrilineal history of its maternal parent species. This qualitative category of outcome (mtDNA paraphyly of a sexual mother species with respect to its clonal derivative) has been reported for the following unisexual taxa: the *Ambystoma* unisexuals (fig. 4.17), *Po. monacha-occidentalis* (Quattro et al., 1992a), triploid biotypes

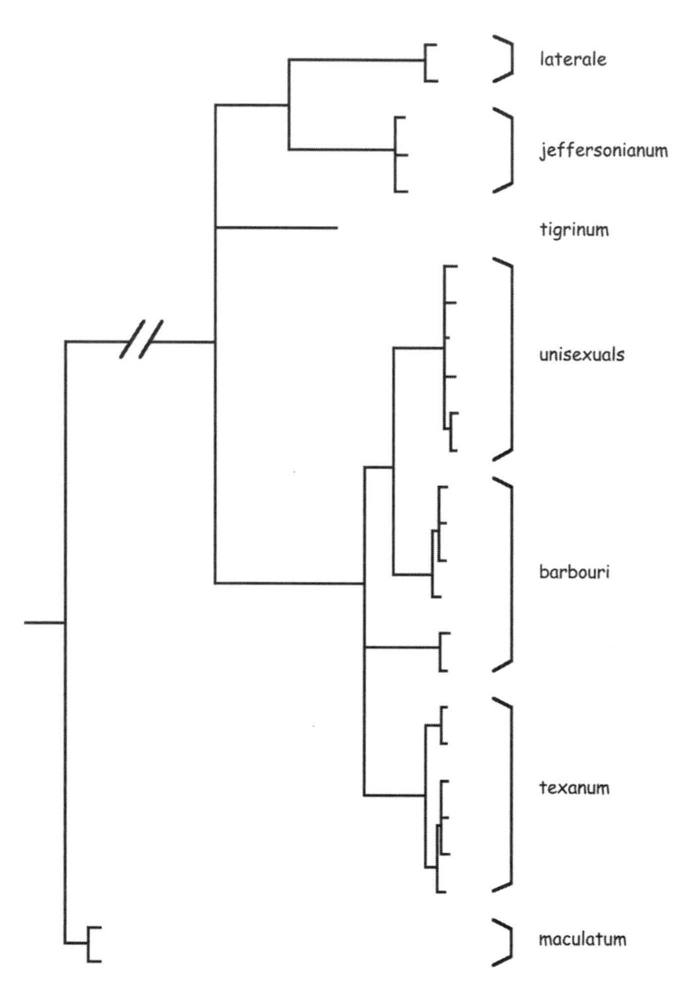

FIGURE 4.17 Phylogenetic tree for mtDNA genotypes observed in natural populations of *Ambystoma* unisexuals and their sexual relatives (after Bogart et al., 2007).

of *Po. monacha-lucida* (Quattro et al., 1992b), and triploid clones of *P. formosa* (Lampert et al., 2005). Such findings suggest severe genetic constraints on either the origin or the long-term evolutionary survival of these unisexual biotypes.

Evolutionary Ages of Clones and Hemiclones

There are several reasons to suspect that gynogenetic or hybridogenetic lineages might sometimes persist for long periods of evolutionary time even if parthenogenetic clones cannot. First, hybridogenetic lineages by definition are hemiclonal rather than clonal, with the paternal components of their nuclear genomes continually rejuvenated by fresh copies in each and every generation. If Muller's ratchet and mutational meltdown are bona fide evolutionary concerns for clonal lineages—and they probably are—then each hybridogenetic lineage presumably should be partially immune to these phenomena because only the maternal genome (the hemiclone) would be fully exposed to the deleterious processes (Leslie and Vrijenhoek, 1980).

Second, some sperm-dependent unisexuals are less than strictly clonal or hemiclonal. For instance, genetic findings for *P. formosa* indicate that small pieces of nuclear DNA from a gynogen's sexual mate sometimes become incorporated into the gynogenetic offspring (Schartl et al., 1995a; Lamatsch et al., 2004; Nanda et al., 2007). Although challenging to document (see Halkett et al., 2005), such low-level paternal leakage of a sexual host's nuclear genes into an otherwise clonal lineage might be a source of adaptive genetic variation, and perhaps offers hope that a gynogen can avoid mutational meltdown under Muller's ratchet. Manfred Schartl and colleagues concluded that "gynogenesis may not be an imperfect parthenogenesis, but a well adapted reproductive mode that combines advantages of sexual and asexual reproduction" (1995a, p. 71). However, Leo Beukeboom and colleagues (1995) criticized this latter interpretation on at least two grounds: it is dubious whether incorporating small segments of DNA could effectively counter Muller's ratchet, and the sexual host upon which the gynogen relies would fail to benefit from such DNA transfer (unless the leaky DNA somehow fed back into the sexual species). This debate merely hints at the broader, and still open, evolutionary question: "Is a little bit of sex as good as a lot?" (Green and Noakes, 1995, p. 87). The question is complicated further by the fact that genetic exchanges may occur along a full continuum. In the case of *Ambystoma* salamanders, evidence suggests that the nuclear genome of a unisexual line is refreshed routinely by genetic recombination with sperm-derived DNA from the sympatric sexual hosts (Bogart et al., 1989, 2007; Kraus and Miyamoto, 1990; Spolsky et al., 1992).

A third reason that some sperm-dependent unisexual lineages might achieve evolutionary longevity applies especially to hybridogenetic forms. In each generation, many new genotypes are created as a result of syngamy between sperm and egg cells. These genotypic assemblies are ephemeral (only the matrilineal component typically passes to the next generation), but some of them might suit their bearers well to the particular environmental conditions of the moment. This may be especially true given that a hybridogen's paternal genes can continually evolve in response to natural selection in the sexual populations where they are otherwise housed. So the paternal genomes that hybridogens borrow (and express) from their sexual hosts may confer these unisexuals with genetic benefits that are unavailable to parthenogens or gynogens. Arthur Bulger and Jack Schultz (1982) uncovered an empirical example: some northern populations of *Po. monacha-occidentalis* profit from a cold-tolerance genome that they borrow, or parasitize, each generation from their sexual host *Po. occidentalis*. In general, functional interactions between the particular maternal (clonal) and paternal (sexual) genomes within each hybridogenetic individual undoubtedly influence the animal's genetic fitness (Wetherington et al., 1989a, 1989b; Semlitsch et al., 1996).

In chapter 3, I described two standard molecular genetic yardsticks (and their inherent limitations) for estimating the evolutionary ages of extant unisexuals: the magnitude of postformational genetic variation within a monophyletic unisexual lineage, and the magnitude of genetic divergence between a monophyletic unisexual and its closest sexual relative. By both of these empirical criteria, most extant parthenogenetic lineages in vertebrates appear to have arisen quite recently in evolution. Indeed, in many cases the same alleles, both nuclear and mitochondrial, that are displayed by extant parthenogens are a subset of those that still segregate within the sexual species from which they arose. Do similar conclusions apply to vertebrate gynogens and hybridogens? Perhaps surprisingly, the provisional answer is yes. Most of these clonal or hemiclonal biotypes have proved to be nearly indistinguishable in mtDNA sequences from their closest sexual relatives, suggesting origination times within (at most) the last 500,000 years (fig. 4.18). Thus, notwithstanding the potential for long-term evolutionary persistence (as described above), sperm-dependent unisexual lineages typically appear to be evolutionarily young, both in sidereal time and in relation to the matrilineal ages of their sexual congeners.

On the other hand, a few unisexual lineages reportedly differ from their closest known sexual relatives by substantial numbers of mtDNA base substitutions, implying (at face value) ancient dates of origination (fig. 4.18). For example, Blair Hedges and colleagues (1992) and Chris Spolsky and colleagues (1992) concluded from such evidence that some unisexual salamander lineages might be as old as five million years, although more recent discoveries of closer extant genetic relatives place the estimate closer to three million years (Bogart, 2003; Bogart et al., 2007). However, if nuclear genomes from sexual host species periodically invade these amphibian lines (as is suspected; see above), then deep lineage antiquity might apply only or primarily to the mtDNA genome per se. Indeed,

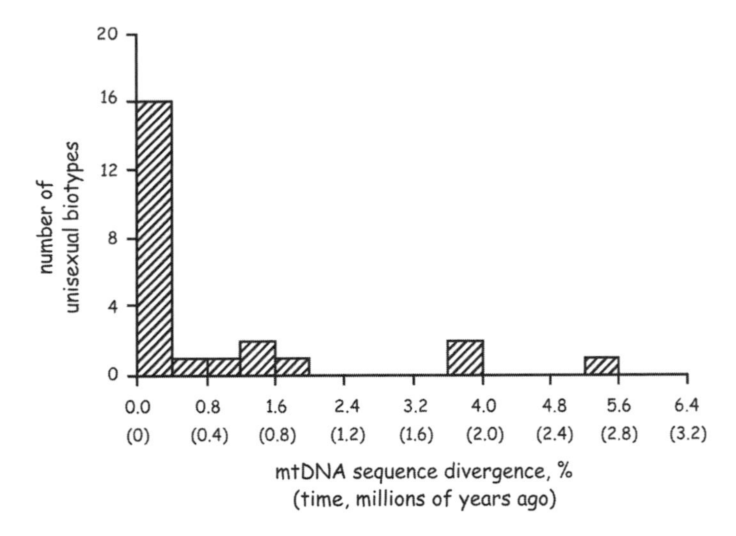

FIGURE 4.18 Histogram of published mtDNA genetic distances for 24 unisexual vertebrate biotypes and the respective extant sexual relatives to which each is genetically closest (after Avise et al., 1992). Also shown is a provisional scale of inferred origination times for the unisexuals, assuming a standard clock calibration for vertebrate mtDNA (2% sequence divergence per million years; Brown et al., 1979). About half of the comparisons in this graph involve parthenogens and the other half involve gynogenetic or hybridogenetic lineages.

the phenomenon of nuclear genome replacement could be key to the apparent persistence and success of these unisexuals.

Quattro and colleagues (1992a) conducted an especially critical study of the evolutionary age of one hybridogenetic vertebrate lineage. The authors reported substantial postformational mtDNA diversity within a monophyletic line of *Po. monacha-occidentalis* that from independent biogeographic evidence, protein-electrophoretic data, and tissue-graft analysis had arisen via a single hybridization event. Based on the evolutionary rate at which mtDNA mutations typically accumulate, Quattro and colleagues (1992a) concluded that this hemiclonal lineage was probably about 60,000 years old, making it among the most ancient unisexual vertebrates with fairly secure genetic documentation. In a commentary about this study, John Maynard Smith rightly noted that 60,000 years "is but an evening gone" in evolutionary time (1992, p. 662) and that the molecular findings for *Poeciliopsis* do not contradict conventional wisdom that most vertebrate clonal lineages are relatively short-lived.

Of course, even if all extant gynogenetic and hybridogenetic lineages were evolutionarily young, this would not imply that the phenomena of sperm-dependent clonality and hemiclonality are themselves recent. Unisexual life-forms (parthenogenetic, gynogenetic, and hybridogenetic) probably have come and gone for at least tens of millions of years. As phrased by Bogart and colleagues, "The combination of the catholic origination of unisexuality among many metazoan lineages with the short temporal existence of such lineages suggests that unisexuality is a constant but pervasive evolutionary process" (2007, p. 120).

Comparative Ecology and Natural History

It was shown in chapter 3 that some parthenogenetic lineages can be highly successful, ecologically, during their relatively brief times on the evolutionary stage. Can the same be said for gynogenetic and hybridogenetic lineages? Like parthenogens, sperm-dependent unisexuals have high within-individual heterozygosity—due to their hybrid origins—that might favor, via heterosis, their survival and reproduction in particular environments. Many sperm-dependent biotypes also have various quasi-sexual mechanisms (as described earlier) that could contribute to their ecological success. Still, sperm-dependent unisexual lineages would seem to lead a fragile existence that could be terminated at any time by extinction of their host species. Gynogenetic and hybridogenetic taxa are indeed caught in an ecological bind—they can neither outcompete, nor escape from, the species that they sexually parasitize.

The necessity of mating with foreign males imposes ecological and behavioral constraints on sperm-dependent unisexuals. Being wedded to their sexual hosts, gynogens and hybridogens lack fertilization assurance and accordingly have a limited potential—compared to parthenogens—to colonize new areas or achieve broad geographic ranges. The constraints of sperm dependency might be especially severe for hybridogens because the paternal genomes that they parasitize must be functionally compatible with the maternal genome in order to produce viable hemiclonal offspring, yet not so genetically compatible as to recombine extensively with the maternal genome and thereby dissolve the hybridogenetic system (Leslie and Vrijenhoek, 1978; Graf and Polls Pelaz, 1989). Gynogens, by contrast, need merely find sperm suitable for stimulating egg divisions (not syngamy), and therefore might be able to exploit a wider variety of sperm sources. *Poecilia* and *Poeciliopsis* gynogens can certainly each employ the sperm from multiple sexual species (Beukeboom and Vrijenhoek, 1998). This cannot be the whole story, however, because the *Poeciliopsis* hybridogens are far more widely distributed than the gynogens (Beukeboom and Vrijenhoek, 1998).

The continuance of any gynogenetic or hybridogenetic lineage also requires that unisexual females remain capable of duping sperm-donor males who presumably are under selection pressure to avoid being duped (Schlupp et al., 1991; Dries, 2003). This coevolutionary game could play out in several ways. Perhaps sexual males readily evolve the sensory capacity to distinguish, and sexually reject, gynogenetic or hybridogenetic females, thereby quickly terminating unisexual lineages. This would place extant clonal or hemiclonal females under strong selection pressure to mimic sexual females in behavior or morphology (McKay, 1971; Lima et al., 1996). Or sexually parasitized males might come to discriminate disproportionately against common gynogenetic or hybridogenetic phenotypes, and rare or newly arisen unisexual clones might tend to be maintained by this frequency-dependent form of mating (Keegan-Rogers, 1984; Keegan-Rogers and Schultz, 1988). Or perhaps males really suffer little loss in fitness by mating indiscriminately, in which case only females might be choosy (Engeler and Reyer, 2001). Another possibility is that sexual males who mate readily with unisexual females also tend to be more successful in mating with homospecific sexual females, in which case selection for mate discrimination by males might also be relaxed (Kawecki, 1988). For example, Ingo Schlupp and colleagues (1994) found that sexual *P. latipinna* females copy the mating preferences of clonal *P. formosa* females, such that potential fitness benefits accrue, indirectly, to *P. latipinna* males who mate with the gynogens. And finally, with respect to the kleptogens, if a unisexual lineage occasionally incorporates all or relevant portions of the genome from sympatric sexual lineages, the evolution of reproductive isolating mechanisms might be forestalled because the sexual males could have difficulty discriminating unisexual females from sexual females.

In general, sperm-dependent unisexuals must walk a narrow line between the risk of performing too poorly to survive in a particular environment, and the risk of doing too well—in the sense of outcompeting their sexual compatriots—and thereby precipitating their own demise (Schlupp, 2005). Ecological and demographic conditions that can lead to stable evolutionary equilibria between unisexuals and their sexual hosts may be quite stringent (Moore and McKay, 1971; Moore, 1975, 1976; Stenseth and Kirkendall, 1985; Hellriegel and Reyer, 2000). However, this inherent evolutionary instability might be ameliorated if sympatric sexual and clonal forms occupy different microniches within their cramped ecological quarters (Vrijenhoek, 1978, 1998; Kirkendall and Stenseth, 1990). Such niche specialization should also promote a sympatric coexistence of multiple clonal lines (Vrijenhoek, 1989b).

These ideas were motivated in large part by the observation that unisexual clones and hemiclones of *Poeciliopsis* fish in nature typically differ from one another, and from their sexual relatives, in a wide variety of physiological and behavioral traits: microhabitat preferences (Vrijenhoek, 1978); foraging behavior, predation efficiency, and diet (Schenck and Vrijenhoek, 1986, 1989; Weeks et al., 1992); growth rate, sexual aggressiveness, and mate choice (Keegan-Rogers and Schultz, 1984, 1988); swimming endurance and survival under stress (Vrijenhoek and Pfeiler, 1997); and thermal tolerance (Bulger and Schultz, 1979, 1982), to name a few. Among artificial (lab-synthesized) hemiclones of *Po. monacha-lucida,* too, significant genetic variance has been uncovered for life-history traits, including brood size (Schultz, 1982), birth size, and growth rate (Wetherington et al., 1989a). In their native arroyos of northwestern Mexico, sexual *Poeciliopsis* females vastly outnumber asexual females when the unisexual forms are monoclonal, but asexual females tend to outnumber sexual females when the unisexuals are polyclonal (Vrijenhoek, 1979), further suggesting that the asexuals collectively exploit more ecological niches than can any single clone. Niche differences among clones or between clonal lineages and their sexual relatives have similarly been noted in some other unisexual vertebrates, including *Leuciscus* minnows (Martins et al., 1998) and *Rana* water frogs (Rist et al., 1997; Negovetic et al., 2000).

The frozen-niche-variation hypothesis (Vrijenhoek, 1984a) posits that hemiclonal and clonal lineages of *Poeciliopsis* arise recurrently during evolution, and that natural selection then winnows these down to a relatively small number that happen to be genetically predisposed to exploit the available food and spatial resources efficiently. Each asexual lineage in effect is frozen into a specific ecological niche that is largely dictated by its idiosyncratic genomic makeup. The net result in nature is often a structured ensemble of asexual biotypes, and their sexual hosts, that subdivide their shared environment (Vrijenhoek, 1979). A plausible corollary is the suggestion that new asexual lineages are more likely to become established, all else being equal, in habitats that might be described as "ecological vacuums," that is, as being species-poor or otherwise low in interspecific competition.

A competing hypothesis is that the ecological success of a clonal or hemiclonal lineage is due to "spontaneous heterosis" that in turn stems from the hybrid genesis of each unisexual biotype (Schultz, 1971, 1977). Under this idea, genome-wide heterozygosity tends to confer an asexual organism with higher fitness, for instance, by elevating its tolerance to ecological stresses or its resistance to parasites or pathogens. As a possible example, hybridogenetic frogs (*R. esculenta*) seem to be more resistant to hypoxic stress than their sexual progenitors (Tunner and Nopp, 1979). A variant of this hypothesis is that unisexual organisms possess "general purpose genotypes" that confer a broader tolerance to spatial or temporal environmental heterogeneity (Parker et al., 1977; Lynch, 1984b), and some data consistent with this model have appeared for hybridogenetic *Rana* frogs (Semlitsch, 1993). However, observations on synthetic hemiclones of *Poeciliopsis* cast doubt on the generality of any such versions of a spontaneous heterosis model. Among 33 hemiclones experimentally synthesized in the laboratory, only 14 survived beyond the fourth generation and none exhibited enhanced genetic fitness relative to sexual strains of *Po. lucida* and *Po. monacha* (Wetherington et al., 1987). Thus, even when a unisexual lineage displays a broader ecological capacity than its sexual counterparts, this could in some cases reflect multiple origins and interclonal selection (as in the frozen-niche model) rather than hybridity or heterosis per se (Vrijenhoek, 1994).

SUMMARY OF PART II

1. Unisexual (all-female) clonal lineages have evolved from sexual lineages on at least 100 independent occasions in various groups of reptiles, amphibians, and fish. These clonal or hemiclonal taxa and their sexual relatives are unsuitable for studying the evolutionary origins of genetic recombination (because sex evolved long before creatures with backbones), but they can offer insights into the ecological and evolutionary forces that sustain the predominance of sexual reproduction in vertebrate taxa even in the face of occasional transitions to self-sustaining clonality.

2. A parthenogenetic female reproduces clonally and without male involvement. The unreduced ova that she produces are genetically identical to her and to one another. Without any participation by sperm, each egg begins to divide mitotically during the development of next-generation progeny. A gynogenetic female likewise reproduces clonally, except that sperm is required to initiate cell division in her unreduced ova. The sperm cell does not actually fertilize the clonal egg, but merely stimulates it to begin dividing.

3. Hybridogenesis is a "hemiclonal" form of reproduction with elements of both clonality and sexuality. A hybridogenetic female produces reduced (haploid) eggs that carry only the chromosome set that she received from her mother. Each egg is fertilized by a sperm cell, thereby reestablishing the diploid condition in the resulting offspring. The exclusion of the paternally derived set of chromosomes during oogenesis means that a male can be a genetic father but not a genetic grandfather. The intact (i.e., nonrecombined) set of maternal chromosomes is the clonal component of this hemiclonal system.

4. Not all (and indeed perhaps few) sperm-dependent unisexual biotypes are strictly clonal or hemiclonal. Some, most notably in *Ambystoma* salamanders, incorporate nuclear DNA from related sexual species quite routinely, and various others probably do so at least occasionally. For such unisexual biotypes, the term "kleptogenetic"—implying theft of foreign genetic material—recently has been suggested.

5. The sperm that are utilized by gynogenetic and hybridogenetic females typically come from males of one or more related bisexual species. Such males are said to be "sexually parasitized" because they gain no lasting genetic fitness benefits by "donating" some of their sperm to the unisexual taxon. The obligate

involvement of heterospecific males in gynogenetic and hybridogenetic reproduction places additional ecological, behavioral, distributional, and coevolutionary constraints on sperm-dependent unisexual lineages beyond those that are faced by their sperm-independent parthenogenetic counterparts.

6. Close inspections of nuclear and mitochondrial markers have revealed several details about the cellular and genetic mechanics underlying parthenogenetic, gynogenetic, and hybridogenetic reproduction, including how various polyploid as well as diploid unisexuals likely were produced. Such cytonuclear analyses have also confirmed that all known unisexual vertebrate taxa arose in evolution via hybridization between related sexual species, and in numerous cases they have enabled specification of the particular parent species, direction(s) of cross, and estimated numbers of formational hybrid events.

7. Molecular genetic data have revealed that all known clonal and hemiclonal vertebrate lineages that are alive today arose rather recently in evolution and thus occupy only the outermost twigs of the vertebrate phylogenetic tree. The maximum well-documented age of any extant vertebrate clonal line is about 60,000 generations, although the possibility cannot be excluded that a few such unisexual lineages might be as much as a few million years old. In any event, no unisexual lineage has adaptively radiated into multiple taxonomic species or otherwise participated appreciably in the macroevolution of vertebrate clades. As reproductive modes, however, clonality and hemiclonality probably have arisen throughout much of the evolutionary history of fish, reptiles, and amphibians, with each unisexual lineage likely going extinct soon after its origin.

8. Despite their relatively short durations on the evolutionary stage, some unisexual vertebrate lineages have achieved remarkable ecological success. Whereas some such taxa are rare and localized, others are abundant (in absolute numbers or compared to related sexual species), occupy multiple niches, and/or have achieved broad geographic distributions.

9. Unisexual clonality in vertebrates can thus be viewed as a shortsighted genetic operation that is sometimes highly successful in the ecological short term but that almost invariably fails as a long-term evolutionary strategy. Unisexual clonality also illustrates the happenstance nature of evolutionary processes. Each clonal lineage arose spontaneously via interspecific hybridization, rather than gradually by long-term selective processes. The genomes from two (or more) species that were unceremoniously thrown together in the genesis of each unisexual biotype clearly disrupt normal meiotic and sexual operations, yet they sometimes collaborate well enough to yield viable progeny and ecologically successful clonal or hemiclonal lines.

PART III

Sexual Clonality in Nature

"Sexual clonality" might seem like an oxymoron, but in fact two routes exist by which genetically identical vertebrate individuals can arise from sexual crosses involving the union of sperm and egg. Under polyembryony or monozygotic "twinning" (chapter 5), two-parent offspring within a brood are clonemates by virtue of stemming from a single fertilized egg. Under intense multigeneration inbreeding (chapter 6), individuals within a sexual lineage may become so nearly identical in genetic composition as to be, in essence, clonemates. These two forms of sex-based clonality differ from each other, and from the various types of unisexual vertebrate clonality discussed in part II, in several important genetic, ecological, and evolutionary ways. These differing attributes are the topic of part III.

CHAPTER FIVE

Clonality in Utero: Polyembryony

According to East Indian mythology, the princess Gandhara gave birth, after two years of pregnancy, to a piece of flesh that was then divided into 101 portions, 100 of which developed into sons and one into a daughter (Newby, 1966). As noted by Ian Hardy (1995a), this fanciful tale highlights some real-life features of genuine polyembryony: the emergence of multiple offspring from a single genetic source, and a highly skewed sex ratio within the clutch.

In modern human populations, approximately 1% of successful pregnancies result in the birth of twins; in about one-third of those cases the twins are genetically identical, that is, they are clonemates (Bulmer, 1970; MacGillivray et al., 1988). Identical twins, which invariably are of the same sex (barring developmental anomalies), arise when a single fertilized egg—or a small collection of mitotically generated cells derived therefrom—divides into two before embryonic development (embryogenesis) ensues within the mother's uterus. Thus the twins in each such monozygotic set fully share a unique combination of paternal and maternal genes that had been united in a gametic matrimony between one sperm cell and one oocyte. Monozygotic or identical twins are to be distinguished from dizygotic or nonidentical twins, which reflect cellular unions between two separate pairs of sperm and egg. In terms of genetic makeup, dizygotic twins are as different from one another as are full siblings from separate pregnancies.

Polyembryony is the scientific term for "twinning"—the production of genetically identical offspring within a pregnancy or clutch. It is an intragenerational rather than intergenerational form of clonality. Polyembryonic broods need not be confined to clonemate pairs. For example, monozygotic quadruplets constitute each litter of the nine-banded armadillo; similarly, in some species of parasitic hymenopteran insect, hundreds of clonemate offspring may emerge from a caterpillar host into which a female wasp had laid a fertilized egg (Godfray, 1994; Ode and Strand, 1995). In humans, rare instances of monozygotic triplets (1 in 50,000 births) and even monozygotic quadruplets are also known (Markovic and Trisovic, 1979; Dallapiccola et al., 1985; Steinman, 1998).

Monozygotic siblings can be of special interest for the information they provide about genetic versus environmental impacts on various phenotypic traits (box 5.1). The phenomenon of polyembryony is also scientifically interesting because it seems to be an evolutionary paradox (Craig et al., 1995, 1997). Why would natural selection *ever* favor the production of clonemate polyembryos, as opposed to genetically diverse offspring, in a clutch? George Williams (1975) analogized the phenomenon to the purchase of multiple lottery tickets with the same number, even though no reason exists, a priori, to prefer one number to another. In polyembryony, the parents' entire evolutionary wager for each litter is placed on just one photocopied genotype; furthermore, because that genotype was sexually generated and thus differs from those of both parents, at the outset it is functionally untested and ecologically unproven (unlike the intergenerational clones that are perpetuated by parthenogenesis). So the evolutionary paradox is that polyembryony appears to lack the fitness advantages traditionally associated with either sexual or parthenogenetic reproduction. Instead, polyembryony seems at face value to combine the worst elements of sexuality and clonality.

The Cast of Players

The recognition and scientific analysis of polyembryony has a surprisingly deep history (box 5.2). Indeed, as gauged by a perusal of the scientific literature, "per capita" interest in the phenomenon was perhaps greater in the late nineteenth and early twentieth centuries than in any subsequent decade.

Sporadic Polyembryony

As a rare or sporadic occurrence within a given species, polyembryony is taxonomically widespread. Known or suspected monozygotic twins have been recorded occasionally in diverse vertebrates, including cattle (Ensminger, 1980), pigs (Ashworth et al., 1998), deer (Robinette et al., 1977), whales (Zinchenko and Ivashin, 1987), various avian species (Berger, 1953; Olsen, 1962; Pattee et al., 1984), and fish (Laale, 1984; Owusu-Frimpong and Hargreaves, 2000). The phenomenon, nevertheless, is probably greatly underreported, because to my knowledge

BOX 5.1 Twin Studies in Nature/Nurture Debates

Because of their clonal makeup, polyembryonic littermates are useful for assessing developmental or environmental contributions to interindividual variation in particular phenotypic traits. For example, in several physical and metabolic features, armadillo clonemates show significant variation that must be due to environmental influences during their development (Storrs and Williams, 1968). Nevertheless, within-litter variability for most such traits remains significantly lower than between-litter variability (Bagatto et al., 2000), suggesting that the phenotypic features examined may have appreciable genetic components as well.

For humans in particular, three types of twin protocol have been used in scientific studies of genetic versus environmental influences on phenotypes. The least critical approach analyzes trait correlations between monozygotic twins reared apart (MZA twins). At face value, traits that are similar (i.e., positively correlated) in MZA twins seem to suggest the impacts of shared genes on those traits. However, no such finding can fully eliminate the possibility of substantial environmental influences, for at least two reasons: all twins, including those reared apart since birth, shared a prenatal uterine environment that could have had key developmental impacts on the features in question; and relevant similarities may exist also in the twins' postpartum rearing environments, such as if their adoptive families had comparable social or economic status (as is often true).

A second category of twin studies attempts to ameliorate these difficulties by comparing trait correlations for monozygotic versus dizygotic twins, the latter usually matched for sex to avoid biases associated with sex. For genetically hardwired traits, higher correlations are predicted for the monozygotic twin sets. One caveat, however, is that environmental influences to which the monozygotic twins were exposed may have been more similar than those for dizygotic twins, in which case all nature/nurture bets are off. A third variation in twin studies involves comparing monozygotic twin sets reared together versus those reared apart, the rationale being that any relevant environmental effects on the particular features surveyed should result in greater trait differences between the separated twins.

Despite their inherent limitations, extensive twin surveys conducted across several decades have strongly suggested that genes, as well as environmental influences and gene-by-environment interactions, contribute significantly to variation in numerous human phenotypes. Traits documented to have at least a partial genetic basis range from various physical and health conditions to particular behaviors and personality traits, and even to intelligence as measured in standard IQ tests (see, e.g., Loehlin and Nichols, 1976; Bouchard et al., 1990; review in Avise, 1998).

BOX 5.2 Scientific Discoveries about Polyembryony Began in the Late 1800s

The scientific literature on animal polyembryony was recently reviewed by Sean Craig and colleagues (1997), from which some of the following historical information was retrieved.

In 1890 and 1893, S. F. Harmer published pioneering accounts of polyembryony in the Bryozoa, and reports soon followed of polyembryonic reproduction in several other invertebrate taxonomic orders, including Platyhelminthes (e.g., Katheriner, 1904), Cnidaria (Bigelow, 1909), and Echinodermata (Mortensen, 1921). For the Arthropoda (insects and allies), polyembryony involving parasitoid wasps was first described by Paul Marchal in 1898, and a review on that topic was published two decades later (Gatenby, 1918). Other early reviews on polyembryony and sex determination appeared in the journal *Science* (Howard, 1906; Riley, 1907), and in 1927 J. Thomas Patterson provided a comprehensive overview of animal polyembryony that included references to more than 120 relevant scientific articles already available by that time. These and other such accomplishments near the dawn of the twentieth century reflect the breadth and depth of scientific interest in descriptive comparative embryology during that period.

Constitutive polyembryony in vertebrates was also discovered during this era. Hermann von Jhering (1885, 1886) was the first to voice suspicions that the phenomenon was common in armadillos, based on observations that littermates were encased in a single chorion and were invariably of the same sex; about two decades later, Miguel Fernandez (1909) and H. H. Newman and J. T. Patterson (1909) independently published apparent confirmations of that notion. By the time Patterson published his seminal review of animal polyembryony in 1927, he was able to cite 25 published studies that already had dealt with the polyembryony phenomenon in armadillos.

no one has searched methodically—using suitable nuclear genetic markers—for vertebrate polyembryos in nature. In any event, sporadic polyembryony will remain a scientific curiosity more than an evolutionary paradox given that, arguably, no selective explanation need be invoked when only rare and happenstance twins are produced in an otherwise nonpolyembryonic species.

Of greater conceptual interest is any situation in which polyembryony is a common or even constitutive phenomenon in a particular species or taxonomic group. In such cases, there must have been a selective advantage (or at least no insuperable disadvantage) for polyembryony despite its supposed "same-number-lottery-ticket" shortfall. This chapter will focus primarily on constitutive polyembryony in vertebrates, but first some relevant background will be provided on the taxonomic distribution of routine polyembryony in invertebrate animals.

Habitual Polyembryony in Invertebrates

Polyembryony is a common or regular occurrence in at least 16 genera distributed across five invertebrate phyla (Craig et al., 1997): Bryozoa, Cnidaria, Echinodermata, Platyhelminthes, and Arthropoda. For example, bryozoans in the order Cyclostomata are small, colonial marine animals that produce up to hundreds of polyembryonic progeny in specialized brood chambers known as gonozooids (Ryland, 1970; Reed, 1991). In a tiny freshwater hydrozoan (*Polypodium hydriforme;* Cnidaria), polyembryos are generated inside the fish eggs that this parasite infests (Raikova, 1980). And in various species of seastar (Echinodermata, class Asteroidea), larval clones are sometimes budded off from the tips of an adult's arms (Bosch et al., 1989; Jaeckle, 1994).

Some of the manifestations of polyembryony are especially bizarre. In a hermaphroditic flatworm, *Gyrodactylus elegans,* an egg that has been fertilized during an outcross event begins to grow and divide within the worm's uterus. The small assemblage of mitotic cells soon divides unequally, generating a second embryo that starts to develop inside the first, but not until the first (daughter embryo) is released from the parent. A third embryo then begins to develop within the second, a fourth within the third, and so on, eventually yielding as many as 2,500 polyembryonic progeny in four weeks (Baer and Euzet, 1961, as cited in Craig et al., 1997). This Russian-doll configuration of clonemates may be unique to these fish-parasitic flatworms, but it illustrates the lengths to which evolution can go in generating polyembryonic offspring.

A tiny wasp—*Copidosoma floridanum* (fig. 5.1)—that parasitizes moths in the family Noctuidae (Strand, 1989a, 1989b) illustrates another fascinating expression of polyembryony. A female wasp oviposits one or two eggs into an egg of the host moth. After the host egg hatches and begins to develop into a caterpillar, the wasp egg divides mitotically and initiates the production of hundreds or even thousands of polyembryos within the host (Grbic et al., 1998). A few of these clonemate wasps become soldiers—with large mandibles—that patrol inside the caterpillar's body to prevent subsequent invasion by other parasitoids (Cruz, 1981; Hughes, 1989; Giron et al., 2004). Other members of the polyembryonic brood eventually kill the caterpillar by eating their way out of its body. The wasp larvae then pupate on the corpse's skin. *Copidosoma floridanum* is just one of many polyembryonic wasp species distributed across four distantly related hymenopteran families: Encyrtidae, Platygastridae, Braconidae, and Dryinidae. Polyembryony thus appears to have evolved in these parasitoids on at least several independent occasions (Craig et al., 1997).

FIGURE 5.1 *Copidosoma floridanum,* a parasitoid wasp.

More than two-thirds of the independently evolved instances of polyembryony across the invertebrates involve creatures with parasitic lifestyles. Indeed, endoparasitism—in which the parasite resides inside a host's body for part of its life cycle, as illustrated by the hymenopteran parasitoids—predominates over ectoparasitism (in which a parasite remains on the outside of a host). However, cases of the latter are also known (e.g., the *Gyrodactylus* flatworms described above). Sean Craig and colleagues (1997) interpreted the associations between polyembryony and parasitism as evolutionary responses to selection pressures related to a female parasite's inability to predict her optimal brood size (see beyond), rather than to some other ecological peculiarity of parasitism.

Constitutive Polyembryony in Vertebrates

Long-nosed armadillos in the genus *Dasypus* are the only vertebrate animals known to produce polyembryonic litters consistently, and apparently exclusively. The six extant species are primarily South American, but one—the nine-banded armadillo, *Da. novemcinctus* (fig. 5.2)—has extended its range from South and Central America into the southern United States during the last century (Taulman and Robbins, 1996). Within its broad geographic distribution (fig. 5.3), this species can also achieve high local abundances. However, not all *Dasypus* species currently enjoy such ecological success. In particular, *Da. pilosus* is known only from the southwestern Peruvian Andes and is listed by the World Conservation Union as vulnerable to extinction.

FIGURE 5.2 *Dasypus novemcinctus,* the nine-banded armadillo.

FIGURE 5.3 Geographic range of the nine-banded armadillo.

A typical litter of *Da. novemcinctus* consists of genetically identical quadruplets, although unusual instances of twins, triplets, quintuplets, and sextuplets have been reported (Newman, 1913; Buchanan, 1957; Galbreath, 1985). *Dasypus sabanicola* and *Da. septemcinctus* usually produce litters of four or eight. In other species of the genus, standard litter sizes vary from as few as two (in *Da. kappleri*) to eight or nine (*Da. hybridus*), occasionally even reaching twelve.

Polyembryony in *Dasypus* armadillos was first suspected from indirect field and laboratory evidence: littermates seemed invariably to be of the same sex,

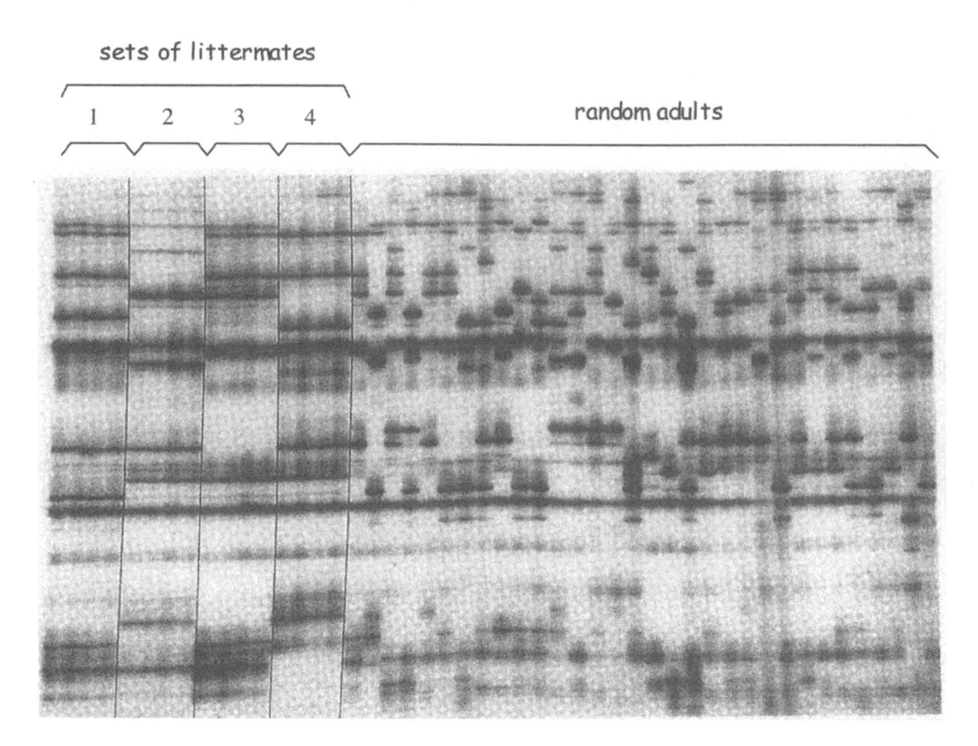

FIGURE 5.4 DNA fingerprints for nine-banded armadillos (after Prodöhl et al., 1996). Note the identical DNA profiles for progeny within each of four surveyed litters (leftmost 16 lanes of the multilocus microsatellite gel), contrasted with the obvious diversity in DNA banding patterns in presumably unrelated adults (32 lanes on the right side of the gel).

and each pregnant female's embryos were encased in a single chorionic membrane (box 5.2). Not until the 1990s, however, were suitable molecular markers finally employed to document, more directly, the clonal identity of littermate progeny (Prodöhl et al., 1996, 1998; fig. 5.4). If any doubts persisted, a recent study demonstrated that armadillo quadruplets accept artificial skin grafts from one another (Billingham and Neaves, 2005), thus indicating a lack of allelic differences at otherwise highly polymorphic histocompatibility loci (see box 4.2).

Also in the family Dasypodidae are several other armadillo genera. The closest relatives of *Dasypus* may be *Tolypeutes* and *Cabassous,* whose species usually produce only one offspring per pregnancy (Wetzel, 1985). This suggests that polyembryony in the *Dasypus* lineage likely arose from an ancestral condition of one pup per litter.

Genetic and Embryological Mechanisms

The genetics of polyembryony are straightforward: a zygote divides mitotically, at least once, and then the two or more derivative cells or sets of cells each initiate the development of a separate embryo. Because only mitotic cell divisions are involved throughout the operation, any resulting littermates are also clonemates.

The embryology of the process, as seen in armadillos, is more intriguing. Early descriptions of embryogenesis in the nine-banded armadillo (Newman and Patterson, 1910; Patterson, 1913; Hamlett, 1933) revealed two developmental peculiarities that researchers have speculated might have causal connections with polyembryony. The first peculiarity is delayed implantation, in which the blastocyst—a postzygotic mass of cells—undergoes a quiescent period of suspended animation (embryonic diapause) before implanting into the female's uterus. Armadillos typically mate in the summer, but in each female the blastocyst from the single resulting zygote does not implant into the uterine wall until months later, in the autumn. A litter of clonemates is born the following spring. These observations led to early speculation that delayed implantation itself might somchow cause polyembryony, perhaps by starving the blastocyst of oxygen (Stockard, 1921a, 1921b) or otherwise altering its physiology in ways that might prompt polyembryonic divisions (Newman, 1923). Delayed implantation also occurs, however, in several other mammals that are not polyembryonic, including some seals, bears, skunks, and weasels (Renfree, 1978; Mead, 1989; Avise, 2006, pp. 110–113). Thus, as first pointed out by G. W. D. Hamlett (1933), the association of armadillo polyembryony with delayed implantation might be spurious rather than causal.

The second peculiarity of armadillo pregnancy is the female's oddly shaped uterus, which appears to have only one blastocyst implantation site. The clonemate progeny that a pregnant mother eventually delivers therefore arise from polyembryonic divisions that take place within her uterus, following implantation. As detailed later, this physical "bottleneck" of a single uterine implantation site might play a key role as a proximate causal agent in armadillo polyembryony.

Ecology and Evolution

> *In discussions on the subject of polyembryony it has not been customary to emphasize the obvious fact that specific polyembryony is a developmental characteristic which is definitely inherited, and as such it must arise and become incorporated in the hereditary mechanism in a manner similar to that of any other heritable character.*

> *That is to say, it must arise as a variation having survival value, and hence effective in adaptation.* (Patterson, 1927)

Background Theory

The genotypic monotony of polyembryonic offspring within a brood stands in sharp contrast to the genotypic heterogeneity among siblings in a non-polyembryonic brood, and the genotypic distinctiveness of polyembryonic offspring from their parents stands in sharp contrast to the genotypic identity of parthenogenetic progeny to their mother. Therefore, any advantages associated with the production of genetically diverse progeny under sexual reproduction, or with the perpetuation of selection-tested genotypes under asexual reproduction, are both absent within a polyembryonic clutch. Ergo the long-standing scientific quandary: Why would parents in any species routinely produce litters consisting of intragenerational clonemates? Yet the fact that constitutive polyembryony has arisen on several independent evolutionary occasions in diverse animal lineages suggests that selective advantages for the phenomenon must exist under some biological circumstances.

In sexual species, parents who produce nonclonemate as opposed to polyembryonic offspring in effect are hedging their reproductive bets by securing multiple tickets in the genetic lottery. But from each offspring's selfish genetic perspective, a heavy parental investment in one specific genotype (its own, either personally or in its polyembryonically duplicated form) might often be a preferred option. In general, especially in sexual species with extended parental care of offspring, parents and their progeny typically have inherent conflicts of interest over optimal parental investment tactics (Trivers, 1974; Stamps et al., 1978; Parker, 1985). Under this view, polyembryony might be interpreted as a special case in which the offspring's preferred tactic has won the evolutionary tug-of-war (Williams, 1975). The broader theoretical challenge would then be to elucidate specific ecological or ontogenetic conditions under which a genetic disposition for monozygotic twinning can successfully invade the gene pool of a population otherwise engaged in traditional sexual reproduction (Gleeson et al., 1994). On the other hand, the tug-of-war metaphor between parents and offspring may be entirely inappropriate in other biological settings—especially in species lacking extended parental care—because in some circumstances polyembryony might be to the benefit of mean genetic fitness in mothers and offspring alike.

Such musings about optimal brood composition have given rise to several hypotheses for how constitutive polyembryony might have evolved in particular animal taxa. In terms of life-history features, conventional speculation has been that polyembryony might be favored (all else being equal) when early stages of embryonic development are lengthy, when brooding females face unpredictable resource availabilities, and/or when sperm cells are severely limited. All of these conditions may apply to the polyembryonic bryozoans, for example. These species have long brooding periods, their food source (phytoplankton) is patchily distributed, and the brooded eggs are fertilized by water-borne sperm from separate colonies (Hughes et al., 2005) such that the gamete-diluting effects of open seawater (Levitan and Petersen, 1995) could make sperm a limiting resource in bryozoan sexual reproduction (Ryland, 1996).

Another general line of speculation is that polyembryony might tend to evolve when parents have less information about optimal clutch size than do their offspring (Craig et al., 1997). When progeny are in the best position to judge the quality of environmental resources available to them, polyembryony could perhaps allow such offspring to adjust the extent of their clonal proliferation accordingly. This hypothesis was motivated by the observed tendency in invertebrate animals for polyembryony to be associated with parasitism, and in particular with endoparasitism (Godfray, 1994).

Consider, for instance, the endoparasitic hymenopterans. To maximize her fitness, each female wasp who oviposits into the egg of a host would ideally "want to know" how big and juicy the moth caterpillar (her progeny's food source) will eventually become, so that she could properly adjust the number of fertilized eggs that she deposits. But such information about the future is presumably unavailable to her. In any event, the moth egg also offers little space for a female wasp to oviposit a multiegg clutch. Under these circumstances, both the mother wasp and her progeny should benefit in terms of expected genetic fitness if proliferation within the brood was delayed until the caterpillar stage of the host. Polyembryony is the only available postoviposition mechanism by which such proliferation can be achieved. A further exigency for polyembryony in this situation is that multiple offspring presumably are necessary to patrol and defend the caterpillar's interior against further parasitic invasions, and eventually to eat their way out of the caterpillar host. And a further advantage from polyembryony in this setting is that competition among clonemates should be minimized, and any collaborative behaviors generally rewarded, under the intense kin selection (see beyond) that presumably should come into play in the tight ecological quarters of a caterpillar host (Hardy, 1995b).

A closely related idea is that polyembryony is selectively favored especially when parents are unable to lay enough fertilized eggs to take maximum advantage of available resources in the local environment. Such may be true, for example, when sperm are in limited supply (as in the cyclostome bryozoans mentioned earlier), when there are space constraints at the oviposition sites (as in the parasitic wasps), or when special ecological opportunities for embryos are postponed until well after oviposition (again, as in some of the endoparasitic hymenopterans). As described next, some of these ideas may also help to explain the evolution of polyembryony in armadillos.

The Genesis and Maintenance of Polyembryony in Armadillos

For creatures with internal pregnancy, such as armadillos, polyembryony is especially enigmatic. One might suppose that in any such species with heavy and long-lasting maternal commitments to progeny, a mother would do better to invest her finite reproductive capital in genetically diverse (rather than clonemate) offspring. Yet polyembryony appears to be *the* constitutive reproductive tactic in *Da. novemcinctus* and probably in all *Dasypus* species.

Kin Selection

One hypothesis is that nepotism (kin-favoring behavior) may have played a role in the evolution of armadillo polyembryony. In 1964, the famous evolutionary biologist William Hamilton formally introduced the important notion that an individual in any species can transmit copies of its genes to the next generation in two ways: directly by producing offspring, or indirectly by helping close relatives leave descendants. The former is *personal* genetic fitness, whereas the latter contributes to an individual's *inclusive* genetic fitness. Hamilton demonstrated mathematically that any gene encoding nepotistic behavior can spread in a population if the cost that it entails to the individual (mean loss in personal reproduction) is more than compensated by enhanced transmission of the gene's copies through the nepotist's relatives. Such kin selection should operate far more effectively on close relatives (who likely share copies of any gene in question) than on distant cousins (who may not).

Clonemates are obviously the nearest of kin and, thus, might be especially prone to the evolution of nepotistic behaviors. As applied to armadillos, the kin selection hypothesis posits that polyembryony arose and/or is maintained by fitness benefits that clonal pups presumably experience through mutually favorable interactions. In other words, via nepotistic behaviors, the mean survival and reproduction of clonemate pups is predicted to be higher than would be the case for comparable nonclonemate litters. In theory, littermate nepotism in armadillos might take any of several forms: cooperation in constructing or defending burrows; predator surveillance and warning; cooperative foraging; mutual grooming to remove ectoparasites; or other behaviors that reflect an "all-for-one-and-one-for-all" mentality among the clonal members of a brood.

Do armadillo littermates truly behave nepotistically? Any good opportunities to do so probably would have to take place in the first few months of life, when the pups of a litter often forage together and share a burrow before going their separate ways in the late summer or early fall. For several years, Jim Loughry and Colleen McDonough at Valdosta State University in Georgia have conducted field observations and experiments designed to test whether nine-banded armadillos behave in ways that might be consistent with the kin selection hypothesis for the maintenance of polyembryony. Their data indicate that the animals are not overtly nepotistic.

For instance, in controlled experiments in behavioral arenas, Loughry and colleagues (1998a) found no statistically significant tendency for juvenile littermates to interact more amicably, or even to stay closer to one another, than to nonclonemates, despite these animals' demonstrated capacity to distinguish kin from non-kin by olfactory cues (Loughry and McDonough, 1994). Perhaps nepotism is not expressed until much later in life, if for example territorial adults behave less aggressively toward littermates than non-littermates. However, this, too, seems unlikely for at least two reasons: (1) juvenile mortality is extremely high in nine-banded armadillos (McDonough and Loughry, 1997), so the survival of multiple littermates is unusual; and (2) clonemates that do survive into adulthood tend to be dispersed rather than spatially adjacent (Prodöhl et al., 1996), such that any opportunities for meaningful behavioral interactions would seem to be extremely limited.

Uterine Constraint

If kin selection does not account for polyembryony in armadillos, what does? In 1985, Gary Galbreath at the University of Chicago advanced the intriguing notion that the armadillo's oddly shaped uterus, with a tiny implantation site that can accommodate only one blastocyst, was the ancestral condition for the *Dasypus* lineage, and that polyembryony then evolved under the influence of natural selection as a means by which females (and their young) could increase their genetic representation in the population. The uterine constraint hypothesis seems phylogenetically plausible because other armadillo genera usually have litter sizes of one and display a similar uterine morphology.

If the uterine constraint hypothesis is correct, then the evolution of polyembryony in armadillos bears a striking analogy to the evolution of polyembryony in parasitic hymenopterans (Loughry et al., 1998b). For the parasitoid wasps, a tiny host egg is the resource bottleneck that later expands into a spacious caterpillar whose food-rich body can support the development of multiple polyembryonic parasites. For the armadillos, a tiny implantation site is the resource bottleneck that later expands into a spacious intrauterine environment that can house and nourish multiple clonal embryos. Thus, for wasps and armadillos alike, polyembryony circumvents an imposed but temporary restriction on brood size by capitalizing upon what soon becomes a rich, expanded ecological setting (caterpillar or uterus) for embryonic development. The uterine constraint model does not necessarily exclude the possibility that kin selection and nepotism also play a role in the maintenance of armadillo polyembryony, perhaps by minimizing intrabrood competition within a female's uterus in ways that increase the number of successful live births.

Even if the uterine bottleneck hypothesis has merit, it leaves unanswered the interesting question as to why the uterus in the ancestral armadillo lineage had evolved a single implantation site to begin with. This mystery remains unsolved. In any event, the broader evolutionary point is that phylogenetic constraints, in addition to contemporary selective forces, can have important impacts on reproductive modes.

CHAPTER SIX

Clonality by Incest: Hermaphroditic Self-Fertilization

Hermaphroditos was a son of Hermes and Aphrodite (the Greek gods of male and female sexuality), and he had inherited the beauty of both parents. When he was 15 years old, he visited the well near Halicarnassus. There, the lovely nymph Salmacis was captivated by his handsome presence. She yearned to gain his affections, so one day while Hermaphroditos was bathing in the pool, a naked Salmacis embraced him, covered him with kisses, and prayed to the gods to be united with him forever. Her wish was granted: Hermaphroditos thereafter became part male and part female.

Tiresias—a blind prophet-priest of Zeus—was another type of hermaphrodite, one who switched between male and female. This began when Tiresias chanced upon a pair of copulating snakes and beat them with a stick. Hera, the wife of Zeus, was infuriated, and she punished Tiresias by transforming him into a woman. After seven years as a female, Tiresias again encountered two mating snakes. This time she left the animals alone, and as a reward Hera permitted Tiresias to become male again. (This story has an interesting footnote. One day, Zeus and Hera were arguing about who enjoyed sex more—males or females. They asked the question of Tiresias, who had experienced intercourse both as a man and as a woman. Tiresias replied that women enjoy the act far more. Enraged by this response, Hera instantly struck Tiresias blind. Zeus could not stop this, but he compensated Tiresias by giving him the gift of foresight.)

Hermaphroditism is common in the real world, too, especially in plants and invertebrate animals (Leonard, 2006; see also box 6.1). The phenomenon occurs as a rare or sporadic developmental anomaly in many vertebrate species, including humans (Kim et al., 2002). Of greater evolutionary interest are vertebrate

BOX 6.1 Hermaphroditism in Plants and Invertebrate Animals

In the botanical literature, individual plants that have male and female reproductive organs, and that produce both pollen (male gametes) and ova (female gametes), are sometimes said to be "monoecious." In some cases the sexual organs—stamens and carpels, respectively—are housed jointly on each flower (monoecy in a strict sense); in other cases they occur on separate male and female flowers (monoecy in a more generic sense). Depending on the species and ecological factors, a hermaphroditic plant might reproduce by outcrossing (utilizing gametes from another individual) or by self-fertilization (uniting its own pollen and ova). Monoecious or hermaphroditic species are to be distinguished from dioecious or separate-sex (also referred to as bisexual or gonochoristic) species, in which an individual is male or female but not both. Further complicating matters, in gynodioecious plant species (of which there are many), some individuals are monoecious and others are female. In androdioecious species (of which there are very few), some individuals are monoecious and others are male.

In invertebrate animals, hermaphroditism greatly predominates in several phyla, including Porifera (sponges), Ctenophora (comb jellies), Phoronida (small, wormlike creatures), Chaetognatha (arrowworms), and small sessile colonial animals known as Bryozoa (Ghiselin, 1969). The phenomenon is also well represented in the Platyhelminthes (flatworms), Annelida (segmented worms), and Mollusca, among others (Ghiselin, 1969). Many snails, for example, are simultaneous (or synchronous) hermaphrodites, and depending on the species or ecological setting an individual may reproduce either by outcrossing or by self-fertilization.

species in which, as standard practice, each individual produces male and female gametes from functional testicular and ovarian tissue. This situation characterizes many fish species (Atz, 1965; Breder and Rosen, 1966; Thresher, 1984). In most cases, each individual fish is a "sequential" hermaphrodite—like Tiresias—who begins life as one sex and later switches to the other (box 6.2). And in a few extraordinary fish species, the majority of individuals are "simultaneous" (or "synchronous") hermaphrodites—like Hermaphroditos—who can reproduce both as male and female in the same time period (Helfman et al., 1997). One such example involves hamlets (genus *Hypoplectrus*) of Caribbean coral reefs. When a hamlet fish finds a mate, members of the pair take turns playing male and female spawning roles over the course of several nights (Fischer and Petersen, 1987).

BOX 6.2 Sequential Hermaphroditism in Fish

Among the world's approximately 25,000 extant fish species, approximately 500 (2%) are hermaphroditic (Pauly, 2004, p. 108). Nearly all of these are sequential hermaphrodites, of which two categories are distinguished. The first is protandrous hermaphroditism, or protandry, in which an individual begins its reproductive life as a male (with undeveloped or dormant female gonads) but later converts to a functional female. An example of this phenomenon is provided by anemonefish in the genus *Amphiprion*. They live in small groups, each typically composed of one female, one large reproductive male, and several small nonreproductive males (Randall, 2005). When the female dies or deserts the assemblage, the reproductive male soon becomes a female and the next-largest individual becomes sexually active as the harem's new primary male. The second category of sequential hermaphrodites is protogynous hermaphroditism, or protogyny, in which an individual begins its reproductive life as a female (with undeveloped or dormant male gonads) but later converts to a functional male. This pattern is common in wrasse fish in the family Labridae (Warner, 1975; Warner and Swearer, 1991). In the harem-forming cleaner wrasse (*Labroides dimidiatus*), for example, one large male oversees as many as ten females who are arranged in a precise pecking order. If the male dies or vacates, the alpha female quickly begins courting other females and develops functional testes within two weeks (Robertson, 1972).

Conventional wisdom is that fitness trade-offs related to ecological, behavioral, and life-history traits influence whether protandrous or protogynous hermaphroditism evolves in a given fish species (Ghiselin, 1969; Warner et al., 1975). One general consideration is that sperm are individually small and inexpensive to produce compared to eggs, such that body size typically limits female fecundity far more than it does male fertility. Another general consideration is that most fish have more or less indeterminate growth, meaning that body size can increase throughout life. Thus, if all else were equal, it might behoove an individual to produce sperm when young and small but perhaps to switch to egg production after a larger body size has been attained. Protandry is often interpreted as an outcome of this kind of selection pressure. On the other hand, males in many fish species defend scarce territories or otherwise compete intensely for female access, so only larger specimens might expect high reproductive success as males. In such circumstances, protogyny might tend to be the evolutionary outcome, and indeed protogyny is the most common form of hermaphroditism in fish.

Only one vertebrate species, the mangrove killifish, is known to reproduce routinely as a self-fertilizing synchronous hermaphrodite. Self-fertilization ("selfing") is an extreme form of incest, or sex between close relatives. When selfing is continued generation after generation, highly inbred lines emerge whose members are, in essence, genetically identical. For many, if not most, vertebrate species, even modest inbreeding reduces genetic fitness, and this is one reason why the highly incestuous behavior of mangrove killifish is an evolutionary paradox. Another aspect to the paradox is that any clone that has arisen via multigenerational selfing lacks appreciable heterozygosity (fig. 6.1); thus its members miss any of the fitness benefits that heterozygosity might convey. This differs sharply from the situation in any polyembryonic clone (which has normal heterozygosity due to its outcross origin) or any parthenogenetic clone (which has extremely high heterozygosity due to its hybrid origin via an interspecific cross).

In this chapter, we will focus on the puzzling mode of clonality displayed by the mangrove killifish, comparing it to parthenogenesis and polyembryony

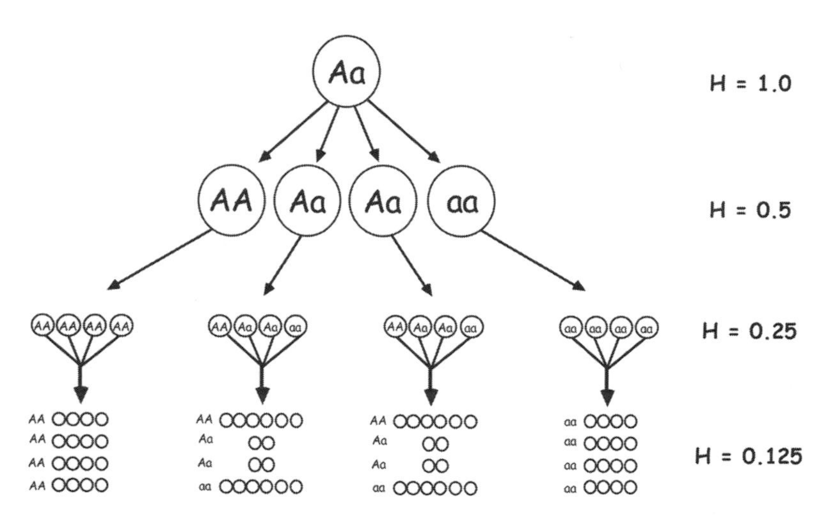

FIGURE 6.1 The rapid loss of heterozygosity (H) under selfing (in this case across four successive generations, from top to bottom). For any autosomal gene that is heterozygous (Aa) initially, a self-fertilizing hermaphrodite produces offspring in an expected Mendelian ratio of 1(AA):2(Aa):1(aa), that is, 50% homozygotes and 50% heterozygotes. Each offspring that is homozygous and self-fertilizes likewise can only produce homozygous progeny, and each offspring that is heterozygous and self-fertilizes again produces homozygous and heterozygous progeny in a 1:1 ratio. The net result is a precipitous expected decline in heterozygosity (by 50% per generation) in any population of self-fertilizing hermaphrodites.

in other vertebrates. We will also examine how the mating tactics of mangrove killifish compare to those employed by various hermaphroditic plants and invertebrate animals.

The Cast of Players

Various levels of inbreeding due to consanguineous mating often characterize natural populations of many vertebrate species (Thornhill, 1993), but rarely, if ever, is inbreeding so extreme and sustained as to yield genetically identical individuals (but see Reeve et al., 1990). The sole evident exception involves the mangrove killifish, *Kryptolebias* (formerly *Rivulus*) *marmoratus*—the only vertebrate known to reproduce routinely by self-fertilization. This tiny fish, only about three centimeters long as an adult (fig. 6.2), has a wide distribution extending from the mangrove forests of peninsular Florida through the Bahamas, most Caribbean islands, the Yucatan Peninsula, and the Atlantic coasts of Central and South America to southeastern Brazil (Davis et al., 1990). It belongs to the killifish family Cyprinodontidae, which has approximately 50 genera and 300 species distributed across the Americas, Africa, and southern Eurasia (Huber, 1992). Although many of these species are poorly known and have not been inspected carefully for reproductive mode, no other cases of selfing hermaphroditism have been discovered to date.

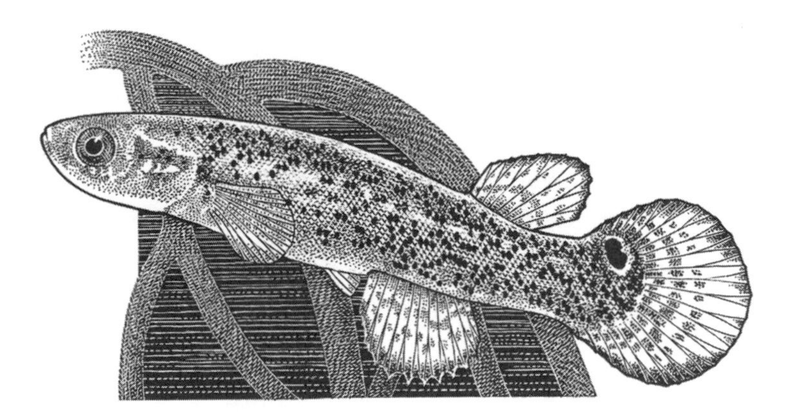

FIGURE 6.2 The mangrove killifish, *Kryptolebias marmoratus*.

Genetic and Reproductive Mechanisms

In *Kr. marmoratus,* each hermaphroditic individual normally fertilizes itself when sperm and eggs that it has produced by an internal organ—the ovotestis—unite inside the fish's body (Harrington, 1963; Sakakura et al., 2006). The gametes arise by the standard meiotic processes of sexual reproduction, except for the important point that genetic recombination in effect is suppressed because each highly inbred strain lacks genetic variation to be shuffled into new allelic combinations. The hermaphrodite then deposits its fertilized eggs in shallow water or in the moist soil of the killifish's coastal mangrove-forest habitat. Embryonic growth, hatching, and juvenile development take place without further parental involvement (Taylor, 1990, 2000).

History

Robert Harrington reported the selfing phenomenon of *Kr. marmoratus* in 1961, and he soon collaborated with Klaus Kallman (see box 4.2) to uncover some of the remarkable genetic ramifications of this peculiar reproductive mode. The researchers found that they could successfully graft fins and organs between a hermaphroditic individual and its offspring, or between progeny within a sibship, thus indicating that these fish were genetically identical and probably homozygous at histocompatibility loci (Kallman and Harrington, 1964; Harrington and Kallman, 1968). By contrast, artificial grafts between some inbred lines were acutely rejected, thus implying that particular selfing strains were genetically different. Harrington and Kallman used the word "clone" to refer to each highly inbred strain of *Kr. marmoratus,* a practice that continues today. This usage is not without shortfalls, however, because "clonemates" could be misconstrued to have arisen via nonmeiotic processes (as in parthenogenesis) and also because the delimitation of a clone can be ambiguous when (as often happens) refined molecular assays uncover cryptic genetic variation within a previously suspected clonal entity.

Each decade since the 1960s has witnessed the introduction of new laboratory methods to assay DNA and proteins more directly at the molecular level (Avise, 2004a). Many of these approaches have been applied to mangrove killifish, including protein-electrophoretic assays (Massaro et al., 1975; Vrijenhoek, 1985), multilocus DNA fingerprinting (Turner et al., 1990, 1992b; Laughlin et al., 1995), mtDNA restriction sites (Weibel et al., 1999) and sequences (Murphy et al., 1999; Lee et al., 2001), molecular surveys of major histocompatibility complex (MHC) loci (Sato et al., 2002), and, most recently, multilocus microsatellite appraisals (Mackiewicz et al., 2006a, 2006b, 2006c). These molecular assays generally have confirmed that natural populations of *Kr. marmoratus* often consist of strains that are so inbred as to be, in effect, clonal. However, some of the assays have revealed far more genetic variation in *Kr. marmoratus* than was formerly appreciated; furthermore, they have identified previously unsuspected genetic and reproductive phenomena in the species (as described in the next section). Table 6.1 illustrates how molecular markers have been employed to confirm the highly inbred nature of *Kr. marmoratus* populations, and also shows how the markers have documented occasional outcrossing between distinctive clones (see beyond).

As illustrated in table 6.2, Floridian populations of *Kr. marmoratus* often show near-zero heterozygosity (within-individual genetic variation) but extensive between-individual genetic variation (clonal diversity). The low heterozygosity clearly is due to the intense inbreeding that accompanies multigenerational selfing. The clonal diversity at particular sites was originally attributed to de novo mutations and interlocality gene flow (Turner et al., 1990; Laughlin et al., 1995); the newer microsatellite data, though, have demonstrated that occasional outcrossing is also a major contributing factor (Mackiewicz et al., 2006a, 2006b, 2006c; Tatarenkov et al., 2007).

The Presence of Males and Outcrossing

Harrington and Kallman (1968) were aware that males do occur at low frequency in *Kr. marmoratus,* and that males can mediate outcross events with hermaphrodites in contrived laboratory settings. This can happen when a hermaphrodite occasionally sheds from its body a few unfertilized eggs, onto which a male (who has no intromittent organ) might release sperm (Harrington, 1963). The males are of two types: (1) secondary males who are hermaphroditic when young but late in life lose ovarian function (Harrington, 1971); and (2) primary males who have functional testicular, but not ovarian, tissue throughout life. Harrington (1967, 1968) discovered that he could readily generate primary males in the laboratory, for example by incubating self-fertilized eggs from a hermaphrodite at low temperature. However, *Kr. marmoratus* males seem to be very uncommon in eastern Florida (where Harrington and Kallman obtained their strains), and this observation—in conjunction with the mostly clonal makeup of natural populations—led the authors to conclude that outcrossing is rare or absent in nature (Kallman and Harrington, 1964).

Males subsequently were uncovered in much higher frequencies (>20%) in some other populations of *Kr. marmoratus* (Davis et al., 1990), notably at Twin Cays in Belize (Turner et al., 1992a, 2006). Initial molecular assays, or DNA fingerprinting, also documented the occurrence of natural outcrossing at the Belize sites (Lubinski et al., 1995; Taylor et al., 2001), and recent genetic reappraisals based on highly polymorphic microsatellite loci suggest that about 55% of the matings in Belize may be outcross events and that 45% involve selfing (Mackiewicz et al., 2006b). Outcrossing has been genetically confirmed in Floridian populations as well (Mackiewicz et al., 2006b, 2006c), albeit at much lower inferred frequencies

TABLE 6.1 Simple example of how molecular data can document clonality, outcrossing, and Mendelian genetics in mangrove killifish[a]

	Locus							
Specimen	*1*	*2*	*3*	*4*	*5*	*6*	*7*	*8*
Clone 1, a	178/178	294/294	214/214	212/212	210/210	166/166	300/300	258/258
Clone 1, b	178/178	294/294	214/214	212/212	210/210	166/166	300/300	258/258
Clone 2, a	158/158	290/290	206/206	224/224	225/225	182/182	284/284	298/298
Outcross	178/158	294/290	214/206	212/224	210/225	166/182	300/284	258/298
Progeny a	158/178	290/290	214/214	212/224	210/225	182/182	300/284	298/298
Progeny b	178/178	290/294	206/214	212/212	210/210	166/166	284/284	298/298
Progeny c	158/158	290/290	206/206	224/224	225/225	166/166	300/300	258/258

[a]Each column is a microsatellite locus, and each row is a different fish specimen (representing two inbred lines, a first-generation outcross progeny between those clonal lines, and three second-generation progeny from the first-generation outcross specimen). The body of the table shows each individual's diploid genotype at each locus; numbers refer to the sizes of different alleles. Observe the complete homozygosity in the two clonal lines, the high heterozygosity in the outcross hybrid, and the intermediate heterozygosity and segregant genotypes in the progeny of the outcross specimen that had self-fertilized. These data are a small subset of information (42 specimens, 36 loci) in Mackiewicz et al. (2006a).

TABLE 6.2 Example of clonal diversity in a Floridian population of mangrove killifish[a]

	Locus							
Specimen	*1*	*2*	*3*	*4*	*5*	*6*	*7*	*8*
a	154/154	310/310	198/198	224/224	219/219	174/174	284/284	278/278
b	154/154	310/310	198/198	224/224	219/219	174/174	284/284	278/278
c	154/154	310/310	198/198	224/224	219/219	174/174	284/284	278/278
d	154/154	310/310	198/198	224/224	219/219	174/174	284/284	278/278
e	162/162	294/294	198/198	224/224	204/204	178/178	284/284	286/286
f	162/162	294/294	198/198	224/224	204/204	178/178	284/284	286/286
g	166/166	298/298	198/198	224/224	228/228	166/166	284/284	278/278
h	134/134	302/302	198/198	224/224	213/213	166/166	284/284	274/274
i	158/158	302/302	198/198	224/224	216/216	174/174	284/284	282/282

[a]Each column is a microsatellite locus, and each row is a different fish specimen. The body of the table shows each individual's diploid genotype at each locus (numbers refer to the sizes of different alleles). These data are a small subset of information (78 specimens, 35 loci) in Mackiewicz et al. (2006b).

(ranging from zero to 20% across ten surveyed locations). Outcross events are often presumed to be male-mediated (Sakakura and Noakes, 2000), because *Kr. marmoratus* hermaphrodites show courtship and spawning behaviors typical of females in other killifish species (Kristensen, 1970) and because male participation has been documented in the laboratory (Mackiewicz et al., 2006a). Still, it remains possible that pairs of hermaphrodites sometimes cross as well.

Another possibility is that some of the outcross events involve crosses between males and young individuals that function solely as females. Using gonadal dissections, Kathleen Cole and David Noakes (1997) found that some relatively young specimens of mangrove killifish are pure females that only later, in adult life, add sperm production to their overall reproductive repertoire. However, it remains unclear whether these young females actually reproduce, or whether they are merely in a transient developmental stage on the path to functional hermaphroditism. Such specimens regardless raise the possibility that simultaneous hermaphroditism in *Kr. marmoratus* might have evolved from an intermediate condition of protogynous (female-first) hermaphroditism.

If we neglect for the moment the possibility of functionally pure females, then the presence of males in addition to hermaphrodites means, by definition, that *Kr. marmoratus* is an androdioecious species (box 6.3) rather than a strictly hermaphroditic species. Furthermore, the presence of outcrossing means that *Kr. marmoratus* actually has a mixed-mating system (box 6.4) rather than one of constitutive self-fertilization. Both of these features, which are unique to known vertebrate animals, merely add to the list of superlatives for *Kr. marmoratus*. All of these biological properties also have important ecological and evolutionary ramifications.

The Fireworks Model

The documentation of at least occasional outcrossing against a backdrop of predominant selfing gives rise to what could be named the "fireworks" model (fig. 6.3) for the population genetic architecture of *Kr. marmoratus* at particular sites such as in Florida. In this metaphor, a black nighttime sky represents the near-complete absence of within-individual heterozygosity in an inbred (highly selfed) population, and each exploding firework represents a single outcross event between distinct homozygous clones. At the core of each explosion is a bright spot of light that represents high heterozygosity in the outcross progeny. Streamers of light, brilliant at first but then quickly fading, burst out of this core as the heterozygous offspring begin to reproduce, often by a resumption of selfing. The many streamers of light that head in different directions represent the many different recombinant genotypes that inevitably arise during this reproductive process, but the streamers fade back into darkness as intrastrain heterozygosity

BOX 6.3 Androdioecy

Biologists have long appreciated that many plant species are gynodioecious—that is, that they consist of mixtures of hermaphrodites and females. Indeed, Charles Darwin discussed the topic at length in an 1877 book. Even a century later, however, there were few, if any, well-documented cases of the apparent converse: androdioecy, or the coexistence of hermaphrodites and males (Pannell, 2002). This paucity of examples was not necessarily surprising; mathematical models suggest that biological conditions favorable for the evolutionary maintenance of androdioecy are highly restrictive (review in Charlesworth, 1984), even more so than for gynodioecy (Charlesworth and Charlesworth, 1978).

Nevertheless, in the last three decades many cases of androdioecy have come to light, specifically in about 50 plant species and 36 animal species, many of the latter being crustaceans (Weeks et al., 2006). From reviews of this scientific literature (Pannell, 2002; Weeks et al., 2006), several empirical generalizations (often tempered by the small number of data points) have emerged:

1. Androdioecy is exceedingly rare in animals, and where it does occur it is usually confined to just one or a few closely related species.
2. The phenomenon usually appears to have evolved from ancestral dioecy (separate sexes), although cases of ancestral hermaphroditism, as in barnacles, are also known.
3. In cases where dioecy was ancestral, androdioecy can frequently be interpreted as an evolved outcome related to selection pressures for the "reproductive assurance" (Baker, 1955) that selfing hermaphrodites enjoy, especially in low-density situations, because of their capacity to reproduce without a partner.
4. The maintenance of androdioecy may be facilitated when a species exists as a metapopulation of small subpopulations that experience frequent extinction and colonization events (Pannell, 1997).

As discussed in the text, several of these considerations probably apply quite nicely to the mangrove killifish.

is rapidly lost in each successive generation of selfing. Then another explosion occurs, perhaps in a different part of the nighttime sky, as an outcross event releases another brilliant but temporary burst of genetic variation available for recombination.

BOX 6.4 Mixed-Mating Systems

Any population of plants or animals that engages in both self-fertilization and outcrossing is said to have a mixed-mating system (Clegg, 1980; Brown, 1989). Much scientific research, using genetic markers, has been devoted to estimating selfing and outcrossing rates—s and t, respectively, where $s + t = 1.0$—in dozens of hermaphroditic species (see, e.g., Schemske and Lande, 1985). A direct approach, multilocus paternity analysis, is applicable when the female parent of each offspring is known, as is often true in seed-bearing plants. Any embryo (inside a seed) that displays alleles other than those carried by its mother must have resulted from an outcross event. However, any offspring that displays only the dam's alleles at every gene was probably also sired by that same hermaphroditic parent. A less direct approach, that of population genetic analysis, can be applied when the dams of progeny are unknown (Hedrick, 2000). If a mixed-mating population is at inbreeding equilibrium with respect to s and t, then the observed heterozygosity (H_{obs}) falls below random-mating expectations (H_{exp}), the inbreeding coefficient becomes $F = (H_{exp} - H_{obs})/(H_{exp})$, and the estimated selfing rate is $s = 2F/(1 + F)$.

Gynodioecious and androdioecious species can also have mixed-mating systems. The outcrossing component is guaranteed (assuming pure females and pure males are reproductively successful), so the behavior of hermaphrodites determines whether mixed mating applies or not. In many monoecious plants, hermaphrodites seldom or never self, for any of several proximate reasons: male and female flowers on an individual may mature at different times; male and female flower parts within a flower may be positioned such that mechanical pollen transfer is unlikely; or self-sterility genes may be present that mechanistically prevent syngamy between pollen and ovules from the same plant. In most hermaphroditic animals, selfing is precluded, if only because sperm and eggs usually are produced at different stages of life (box 6.2). For both plants and animals, conventional wisdom is that mechanisms that inhibit or proscribe selfing have evolved in response to selection pressures stemming from inbreeding depression (box 6.5).

Evolution and Ecology

Constitutive Selfing

At face value, multigenerational self-fertilization would seem to combine the worst features of sexual and asexual reproduction. In standard outcrossing species,

FIGURE 6.3 The fireworks model for mixed-mating systems such as those displayed by *Kryptolebias marmoratus*.

meiosis and syngamy are advantageous in large part because of the recombinant genetic variety they routinely generate. But constitutive self-fertilization stymies the shuffling effects of meiosis and fertilization, because each selfing lineage quickly loses the genetic variation that is otherwise available for meaningful recombination (fig. 6.1). Furthermore, selfing is an extremely intense form of inbreeding, which normally is a costly practice in most species (box 6.5). And in comparison to other forms of clonality, the clones generated by constitutive selfing lack whatever fitness benefits are associated with the high or moderate heterozygosity levels characteristic of parthenogenetic clonality and polyembryonic clonality, respectively (fig. 6.4).

G. Ledyard Stebbins (1957), among others (Grant, 1958; Wyatt, 1988), has argued that pure selfing is an evolutionary dead-end: highly inbred lineages routinely go extinct because of their limited potential for adaptation and speciation. Another suggested (but debatable) factor might be the genetic deterioration, via Muller's ratchet, expected for highly inbred clonal lines (Lynch et al., 1995; Takebayashi and Morrell, 2001). Probably for several, if not all, of these reasons, self-fertilization is seldom observed as a constitutive reproductive mode in nature. Instead, nearly all species of plants and invertebrate animals that contain synchronous hermaphrodites—including gynodioecious and androdioecious taxa—seem to engage in outcrossing either exclusively or at least occasionally as

BOX 6.5 Inbreeding Depression

In populations of most plant and animal species, inbreeding is typically associated with diminished survival or fertility. For example, one literature review found that inbred progeny suffered higher juvenile mortality than outbred progeny in 41 of 44 (93%) of the captive mammal populations surveyed (Ralls and Ballou, 1983). Similar, if not more extreme, outcomes also apply to most wild populations (Frankham et al., 2002). Inbreeding costs can be high—sometimes 50% or more reductions in genetic fitness—but are highly variable in magnitude among species.

Two hypotheses for inbreeding depression have been debated extensively (Charlesworth and Charlesworth, 1987). Under the "dominance" hypothesis, lowered fitness results from particular loci becoming homozygous for rare deleterious recessive alleles that in outbred populations are usually masked in heterozygotes. Under the "overdominance" hypothesis, a genome-wide drop in heterozygosity per se is the causal factor for the fitness decline. Recent literature seems to offer considerable support for the dominance model, at least in plants, with overdominance playing a secondary but still important role (e.g., Carr and Dudash, 2003).

The dominance hypothesis also implies that if a population can survive an initial bout of intense inbreeding, it might thereafter survive indefinitely (or at least until the environment changes beyond the population's genetic scope) because natural selection will have purged its genome of deleterious recessive alleles. By contrast, the overdominance hypothesis predicts that if an inbred population survives, it will continue to perform poorly compared to an outbred population because its heterozygosity remains low. These two hypotheses thus make somewhat different predictions about the evolutionary prospects of populations that in effect are clonal by virtue of the intense inbreeding that accompanies consistent self-fertilization.

part of a mixed-mating strategy. Most populations of *Kr. marmoratus* also engage in occasional outcrossing, even when selfing predominates to the extent that "clonal" lineages routinely arise.

On the other hand, selfing is by definition a component of the mixed-mating systems that are common in plants (Goodwillie et al., 2005) and invertebrate animals (Jarne and Auld, 2006). The evolution of selfing capabilities from outcrossing ancestors is especially common in plants (Grant, 1981; Goodwillie et al., 2005), having occurred, for example, at least 150 independent times in the Onagraceae

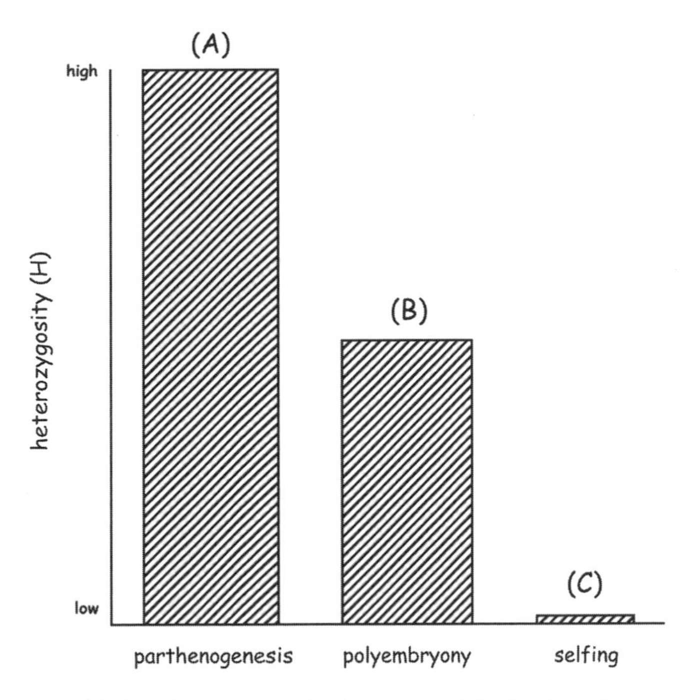

FIGURE 6.4 Heterozygosity levels in clones arising by (A) parthenogenesis, (B) polyembryony, and (C) consistent selfing hermaphroditism. Parthenogenesis is an evolutionary enigma because this reproductive mode dispenses with sex, but at least parthenogenetic clones should retain any fitness advantages that might come from their extremely high heterozygosity due to hybrid origins. Polyembryony is enigmatic because it fails to capitalize fully on the benefits of sexual reproduction, but at least polyembryonic clones should retain any fitness advantages that come from the standard heterozygosity levels that outcrossing provides. Constitutive self-fertilization is especially enigmatic because the clones that it produces lack genetic variation almost entirely.

alone (Raven, 1979). This suggests that selective advantages to individuals often do attend self-fertilization, at least in the short term. One theoretical advantage is intrinsic: a selfer transmits two sets of genes to each offspring, whereas an outcrosser transmits only one set (Nagylaki, 1976; Lloyd, 1979). Other potential benefits from selfing have to do with ecological and behavioral circumstances, as discussed next.

Mixed Mating in Plants and Invertebrate Animals

In many respects, mixed mating potentially converts the worst-of-both-worlds dilemma of constitutive selfing to a best-of-both-worlds adaptive strategy that combines many of the advantages of sexual and clonal reproduction. The history of this notion—which goes back to genetic surveys of hermaphroditic plants and animals in the 1970s—is of interest in its own right.

Coadapted Genomes in Plants

Robert W. Allard was an eminent population geneticist at the University of California at Davis. Beginning in the early 1970s, his laboratory produced a series of seminal papers on the adaptive significance of mixed-mating systems in annual plants (review in Allard, 1975). Much of this empirical work focused on the slender wild oat (*Avena barbata*), a Mediterranean native that was introduced to California by Spanish missionaries about 400 years ago. Each individual oat plant synchronously produces pollen and ova that usually unite in self-fertilization, but that sometimes participate in outcross events.

Allard's group discovered that two highly distinctive multilocus genotypes predominate in Californian populations of the wild oat, one adapted to xeric (dry) conditions and the other to more mesic (wetter) soils (Clegg and Allard, 1972; Hamrick and Allard, 1972). The researchers showed that consistent selfing was often advantageous in the ecological short term because it yields offspring with identical copies of a coadapted multilocus genotype that nature has already field-tested for genetic fitness in a particular habitat. But the researchers also demonstrated that occasional outcrossing was important, too—especially when habitat conditions change over time or show spatial heterogeneity—because outcrossing parents (if from different inbred lines) produce genetically diverse progeny. Natural selection in effect then chooses, from the multitudinous recombinant genotypes produced by outcrossing, particular multilocus combinations of alleles that happen to confer high fitness in the altered regime. A resumption of routine selfing tends to perpetuate these multilocus genotypes intact, until the next round of outcrossing and perhaps novel selection on recombinant genotypes occurs.

Coadapted Genomes in Snails

Robert K. Selander is another eminent population biologist. In the early 1970s, his laboratory at the University of Texas produced an analogous series of studies on the adaptive significance of mixed-mating systems in invertebrate animals (review in Selander and Kaufman, 1975). What Allard demonstrated for monoecious plants, Selander would likewise implicate for hermaphroditic animals.

Most of this research involved land snails, including a hermaphroditic species (*Rumina decollata*) that, like *Av. barbata,* engages in a mixture of facultative self-fertilization and outcrossing.

In its native environment in southern France, two highly distinct inbred strains of *Ru. decollata* predominate, one typically associated with xeric habitats and the other with more mesic conditions (Selander and Hudson, 1976). Yet these "clones" also outcross occasionally, thereby releasing vast stores of genetic variation for subsequent sex-based recombination and selective scrutiny in a given ecological setting. Together with Allard's genetic discoveries on monoecious plants, Selander's genetic findings on hermaphroditic snails provide a powerful example of how the mating systems of plants and animals can adaptively converge in evolution to a mixed-mating strategy that combines favorable elements of both sexuality and clonality.

Baker's Rule

A well-known botanist, Herbert G. Baker, introduced a short paper with the following words: "It is not often that research on an evolutionary topic carried out independently by botanists and zoologists produces conclusions which are virtually identical. When this does happen one cannot restrain from feeling that a principle of more than superficial importance has been uncovered." One might naively suppose that Baker was referring to the works of Allard and Selander cited above, but instead his paper was published two decades earlier (in 1955, p. 347) and dealt with an entirely different feature of potential adaptive relevance for selfing hermaphrodites.

Citing his own research on plants (Baker, 1948, 1953) and A. R. Longhurst's (1955) findings for marine invertebrates, Baker (1955) concluded that the capacities for self-fertilization and for long-distance dispersal are positively correlated across species, and that a plausible explanation involves the reproductive assurance (fertilization insurance) that automatically comes with being a selfing hermaphrodite. In other words, Baker suggested that natural selection has favored self-fertilization capabilities in dispersive species because even a single individual can be a successful colonist. Plants and marine invertebrates that are effective colonizers usually have dispersive propagules—seeds and larvae, respectively—that may be carried for long distances by winds or ocean currents, and upon arrival an immigrant's ability to reproduce self-sufficiently, rather than requiring a partner, is often advantageous if not crucial. The empirical association between self-fertilization and colonizing potential has become known as Baker's rule.

In principle, selfing could be advantageous not just in colonizing species, but whenever individuals experience special difficulties in finding mates. For example, in hermaphroditic species in which population densities are low or individuals are sedentary or solitary, outcross opportunities might be severely limited.

Mixed Mating in the Mangrove Killifish

In principle, microhabitat selection might operate in conjunction with a mixed-mating system to foster the survival and proliferation of particular clonal genotypes in *Kr. marmoratus,* as has been documented in the oat *Av. barbata* and the snail *Ru. decollata.* Few experimental studies of this possibility have been conducted on mangrove killifish (but see Lin and Dunson, 1995), despite their potential feasibility in two key respects: (1) mangrove environments are highly variable in many features, such as salinities, water temperatures, and hydrogen sulfide concentrations, that might impact the fitness of killifish in genotype-specific fashion (Davis et al., 1995); and (2) the fish themselves provide replicate copies of clonal genotypes that can be tested in "common-garden" experiments under controlled, replicated conditions. With the current paucity of such data on genotype-specific selection, there is at present no compelling empirical evidence for or against the proposition that particular multilocus genotypes of *Kr. marmoratus* are uniquely well suited to specific microhabitats in the mangrove environment.

The behavior and natural history of the mangrove killifish can be interpreted, however, as consistent with Baker's rule in several regards.

1. The species has a huge geographic range that extends from southern Brazil to Florida and includes many isolated Caribbean islands. Thus the species clearly is an effective colonizer.
2. The coastal habitats occupied by *Kr. marmoratus,* and especially the tendency of individuals to occupy mangrove litter and termite cavities in rotting logs (Davis et al., 1990), probably predispose this species to occasional long-distance dispersal via floating forest litter, for instance following storms.
3. The dispersive capacity of adults is further enhanced by a life-history feature known as "emersion," wherein an individual can remain out of water for up to ten weeks (Taylor, 1990), respiring and eliminating ammonia cutaneously in damp environments (Grizzle and Thiyagarajah, 1987; Frick and Wright, 2002; Litwiller et al., 2006). Adults can also move overland by flipping, a behavior that helps them escape locally elevated concentrations of hydrogen sulfide, a common toxic component of mangrove habitats (Abel et al., 1987).
4. The fertilized ova are well suited for dispersal because they can survive out of water for prolonged periods, hatching quickly when rehydrated (Ritchie and Davis, 1986). Killifish eggs readily attach to leaves, twigs, or other floating plant debris that often washes to sea, and the embryos are encased in a tough chorion membrane that helps prevent desiccation.

5. Many adult mangrove killifish are suspected to lead rather isolated and independent lives, such as in the tight termite tunnel of a rotting log or the tiny interior of a crab burrow, and they tend to be highly belligerent toward conspecifics.

All of these ecological and behavioral attributes would seem to favor self-fertilization as a routine alternative to outcrossing in *Kr. marmoratus*. Given the mangrove killifish's peculiar habits and natural history, a mixed-mating strategy presumably enables this quasi-clonal species to ameliorate the disadvantages of constitutive outcrossing (the need for mates, and a limited capacity to propagate highly fit multilocus genotypes) as well as those of constitutive clonality (inherent genetic deterioration and inadaptability to changing environmental conditions). Indeed, given that multiple "clonal" strains of the mangrove killifish can certainly escape the perils of intense inbreeding, at least over the short term, this androdioecious species presumably enjoys the best of two worlds: the long-term as well as short-term advantages of outcrossing (continued genetic health and adaptability) *and* the immediate benefits of selfing (fertilization assurance and the intact propagation of locally adapted genotypes).

SUMMARY OF PART III

1. Polyembryony and habitual self-fertilization are two modes by which genetically identical offspring can arise within the standard sexual frameworks of conventional meiosis (production of haploid gametes) and syngamy (fertilization that restores the diploid condition). In the case of polyembryony, two separate individuals (female and male) spawn the gametes that unite to form each zygote, which then divides mitotically and splits before initiating the development of two or more clonemate embryos. In the case of self-fertilization or selfing, the gametes that unite to form each zygote come from a single hermaphroditic individual. When that hermaphrodite is highly inbred, as is often the case, it and its progeny are members of a lineage that in effect is clonal.

2. Polyembryony, or "twinning," is an intragenerational rather than intergenerational form of clonality. It would seem at face value to be an unwise reproductive tactic that can be likened to a reproductive raffle in which parents purchase multiple lottery tickets (progeny) with the identical number (genotype). Nonetheless, the phenomenon is a standard mode of reproduction in a wide diversity of invertebrate animals. In vertebrates, twinning occurs sporadically in various species, but only armadillos in the genus *Dasypus* display the polyembryony phenomenon in essentially every brood.

3. Many polyembryonic invertebrates are endoparasites that spend at least part of their life cycle within a host's body. Wasps that are parasitic on moths provide illuminating examples. In these species, a moth egg is the typical site into which a female wasp deposits an egg that will later divide polyembryonically within the developing host larvae (caterpillar). Polyembryony makes evolutionary sense in this circumstance because the parasite faces an initial but temporary space bottleneck (the host egg) that later will expand into a rich and spacious environment (the caterpillar body) upon which multiple parasitic larvae can feast. These larvae are clonemates, but genetically different from both of their parents.

4. An interpretation analogous to that for the parasitic wasps may apply to the evolution of constitutive polyembryony in *Dasypus* armadillos. In these species, the initial reproductive bottleneck is a peculiarly configured uterus that has only one blastocyst implantation site. Polyembryonic divisions later in pregnancy then give rise to multiple clonemate offspring within the female's eventually enlarged uterus. These littermates are genetically identical to one another but different from both of their parents.

5. For parasitic wasps and armadillos alike, polyembryony can be interpreted as an opportunistic reproductive tactic that may make the best of the available situation for both parental and offspring genetic fitness. In each case, a severe constraint on offspring numbers (small host egg or single site of implantation, respectively) exists at the outset of each reproductive bout, but a more spacious developmental environment (host caterpillar and female uterus, respectively) arises later that can be exploited by multiple polyembryos. Furthermore, interindividual competition among co-housed larvae and embryos should be minimized because the broodmates are also clonemates. If these provisional interpretations about the adaptive significance of polyembryony are correct, they illustrate how ecological idiosyncrasies and phylogenetic legacies can generate selection pressures favoring the evolution of what otherwise might seem to be maladaptive modes of clonal reproduction in sexual outcrossers.

6. Habitual self-fertilization by synchronous hermaphrodites is another sexual route by which genetically identical individuals can arise. However, selfing is an extreme form of inbreeding, even less severe cases of which—from matings between genetic relatives—typically result in inbreeding depression in many plant and animal species. For this and other reasons, constitutive self-fertilization is extremely rare in the biological world.

7. Instead, most species that engage in self-fertilization also outcross at least occasionally, and thus by definition have a mixed-mating system. Lineages in such species can still be transiently "clonal" when selfing rates are high, because self-fertilization continued generation after generation rapidly decreases heterozygosity and thereby squelches any opportunity for meaningful genetic

recombination during meiosis and syngamy. The challenge is to understand the ecological and evolutionary circumstances under which the selfing component of a mixed-mating system arises and persists in otherwise outcrossing species.

8. Two general classes of fitness advantage have been identified for habitual selfers. First is the potential for "clonal" propagation of multilocus genotypes that are coadapted to local habitat conditions. Once homozygosity through inbreeding is achieved, any lineage that has survived the stringent tests of natural selection can perpetuate its genotype, intact, through successive generations of selfing. This benefit of selfing has been documented in some plants and snails that have mixed-mating systems. A second benefit of selfing is fertilization assurance (no need to find a mate for reproduction). "Baker's rule" refers to an observed association in many plants and invertebrate animals between colonization potential and self-fertilization capacity, the interpretation being that individuals in highly dispersive species often find themselves in situations where potential mates are rare or absent.

9. Only one vertebrate species—the mangrove killifish—is known to self-fertilize routinely, and many populations include multiple lineages that are highly homozygous and effectively clonal. The natural history and ecology of *Kr. marmoratus* indicate that this species is a highly effective colonizer, such that its selfing proclivity is entirely consistent with Baker's rule.

10. The mangrove killifish consists of males as well as hermaphrodites, and the species is thus androdioecious. Genetic data have also documented outcrossing between inbred lines, in widely varying frequencies in different populations. This mixed-mating species probably capitalizes jointly on the short-term advantages of selfing (fertilization insurance and the propagation of fit "clonal" genotypes) and both the short- and long-term advantages of outcrossing (genetic health and adaptability).

PART IV

Clonality in the Laboratory

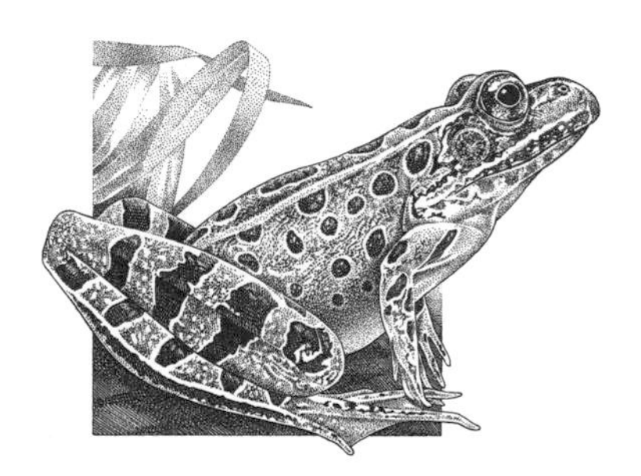

No modern book on vertebrate clonality would be complete without some mention of the "artificial" clones now generated routinely in research laboratories. In recent decades, scientific breakthroughs in molecular and cellular biology, and in reproductive technologies, have given geneticists powerful tools for manipulating DNA and cells directly, including the capacity to create whole-animal vertebrate clones. All of the cloning methods employed by biotechnologists were inspired by, and usually borrowed from, Mother Nature. These laboratory techniques, ranging from gene cloning to organismal cloning, will be the subject of chapter 7.

CHAPTER SEVEN

Human-Sponsored Clonality

Cloning by human hands can mean many things. It can refer to the intentional clonal propagation, via recombinant-DNA technologies, of small bits of a genome such as particular genes or regulatory DNA sequences. It can refer to the artificial stimulation of natural clonal processes, for example by subjecting animals to treatments or research protocols that induce parthenogenesis, gynogenesis, hybridogenesis, or polyembryony, or that enforce intense inbreeding. Or it can mean the purposeful manipulation of cells and whole genomes to generate genetically identical individuals via mechanisms that may differ quite fundamentally from those that occur in nature.

Gene Cloning

In the early 1970s, geneticists discovered how to isolate pieces of DNA from any species, splice the molecules together, and clonally propagate the recombinant molecules by reinserting them into living organisms. In the 1980s, a laboratory technique was invented that permitted scientists to clonally replicate particular pieces of DNA in test tubes. These human-mediated methods of gene cloning—in vivo and in vitro, respectively—are now widely employed in genetic research laboratories and by the biotechnology industry.

In Vivo

Bacteria naturally produce and utilize restriction enzymes that cleave duplex DNA at specific nucleotide positions known as restriction sites. For instance, the human-gut bacterium *Escherichia coli* makes a restriction enzyme (*Eco*RI) that snips duplex DNA wherever the six-letter sequence GAATTC happens to appear; similarly, the bacterium *Bacillus ambofaciens* makes an enzyme (*Bam*HI) that cuts DNA molecules at each GGATCC site. Most restriction enzymes produce staggered cuts that leave short, single-strand terminal ends flanking the resulting snippet of otherwise duplex DNA. Bacterial cells use restriction enzymes in part to destroy ("digest") genetic material of the phage viruses that infect them, so these enzymes can be thought of as biological weapons that help bacteria fend off their microparasites. To avoid self-digestion, each bacterial species also has evolved a chemical "modification" system that protects its own genome from being snipped apart by the restriction enzymes it produces.

All organisms make and use another class of enzymes, known as ligases, that play a natural role in DNA replication. Each ligase catalyzes the formation of tight chemical bonds that can link DNA fragments end to end. Fragments that have single-stranded flanking termini (such as those produced by many restriction enzymes) are especially suitable. A ligase glues such DNA pieces together, much like a cabinetmaker uses glue or screws to solidify the tongue-and-groove joints between dovetailed pieces of wood.

Scientists have purified and commercialized many restriction enzymes and ligases, and these were a primary technical impetus for the recombinant-DNA revolution that began in the 1970s. In a typical genetic engineering project, a biologist uses restriction enzymes and other genetic tools to clip out a locus of interest from the genome of a mammal or any other creature; he or she then uses ligases to attach that gene to another piece of DNA, such as a bacterial plasmid. A plasmid is a tiny, circular piece of DNA that naturally resides and replicates inside a bacterial cell, and a recombinant plasmid is one that has been engineered, as just described, to carry a foreign gene. When a recombinant plasmid is inserted back into a bacterium, it resumes its natural habit of replicating itself, in this case along with the foreign gene that it now carries. Soon a "transgenic" bacterial colony arises in which the foreign DNA has been copied to high numbers (fig. 7.1). This is the usual meaning of "in vivo DNA cloning" in the field of biotechnology.

Two of the first commercial applications of recombinant-DNA cloning involved insulin (the diabetes-treating hormone) and somatotropin (the human growth hormone). In each case, a human gene for the hormone was isolated and inserted into bacteria colonies, which then took over the job of cloning and expressing (making proteins from) the foreign DNA. Soon, large vats of transgenic bacteria

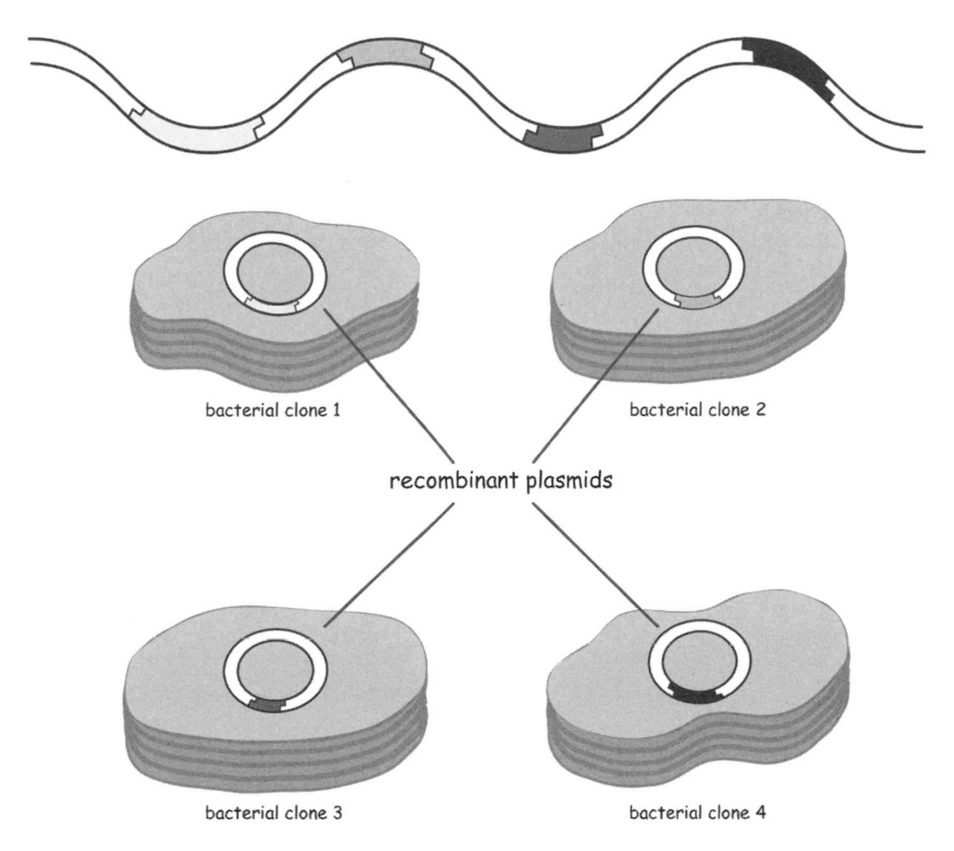

FIGURE 7.1 General outline of one common approach for in vivo gene cloning.

had replaced human cadavers as the primary source for commercial quantities of these medically important compounds. In the last three decades, scientists have engineered transgenic bacteria that bioproduce a plethora of substances of value to mankind, ranging from pharmaceuticals and therapeutic drugs to enzymes with commercial applications in diverse industries such as food processing, cleaning, textiles, paper, leather tanning, and medical diagnostics (Levine, 1999; McGloughlin and Burke, 2000).

Transgenic creatures other than bacteria can also provide gene-cloning services. Indeed, biologists in recent years have produced many kinds of genetically modified organisms (GMOs), ranging from yeasts and other microbial species to macroscopic plants and vertebrate animals. Two examples will suffice here.

Regarding plants, many crops including various strains of corn and cotton have been genetically engineered such that they endogenously produce insecticides (compounds that are toxic to insects), using transgenes that they themselves now clonally propagate through normal processes of DNA replication. Regarding vertebrate animals, transgenic goats have been engineered to carry and express silk-protein genes from spiders. The hope is that the tough silk fibers—which have many potential commercial applications, such as in making bungee cords and bulletproof vests—might be harvested from goat's milk in sufficient quantities to be economically viable. These and dozens of other incredible examples of GMOs in industry, agriculture, animal husbandry, medicine, and the environment are detailed in *The Hope, Hype, and Reality of Genetic Engineering: Remarkable Stories from Agriculture, Industry, Medicine, and the Environment* (Avise, 2004b).

In Vitro

In the 1980s, Kerry Mullis—a scientist then at the biotechnology company Cetus in California—invented the polymerase chain reaction (PCR), an accomplishment for which he received the 1993 Nobel Prize in Chemistry. The PCR is a gene-cloning operation that takes place in test tubes and a benchtop machine, rather than in living organisms. The process begins when two short pieces of artificially synthesized DNA (known as PCR primers) are annealed to nucleotide sequences flanking a particular target gene of interest. The procedure is carried out in such a way that when a special enzyme (*Taq* polymerase) is added to the mix, it catalyzes the synthesis of exact copies of the target gene. The process, outlined in figure 7.2, involves repeated rounds of denaturation of genomic duplex DNA into single strands, primer annealment to the target-flanking strands, and the synthesis or "extension" from those two primers of intervening double-stranded DNA under the direction of *Taq* polymerase. In less than an hour, the PCR molecular chain reaction yields millions of identical copies of the original target locus. Almost any gene from any organism can be clonally "amplified" by this process and thereby made available in sufficient quantity for scientists to manipulate and analyze. The PCR process is employed routinely in laboratories around the world in a vast array of applications, and it would be difficult to overstate its importance to molecular genetic research and the biotechnology industry.

In both the in vitro and in vivo methods of human-orchestrated gene cloning, the genetic engineers are merely arranging special conditions in which to capitalize on nature's own evolved mechanisms and molecular machinery for DNA replication (see chapter 1). In this important sense, the laboratory methods for gene cloning are partly contrived and partly natural.

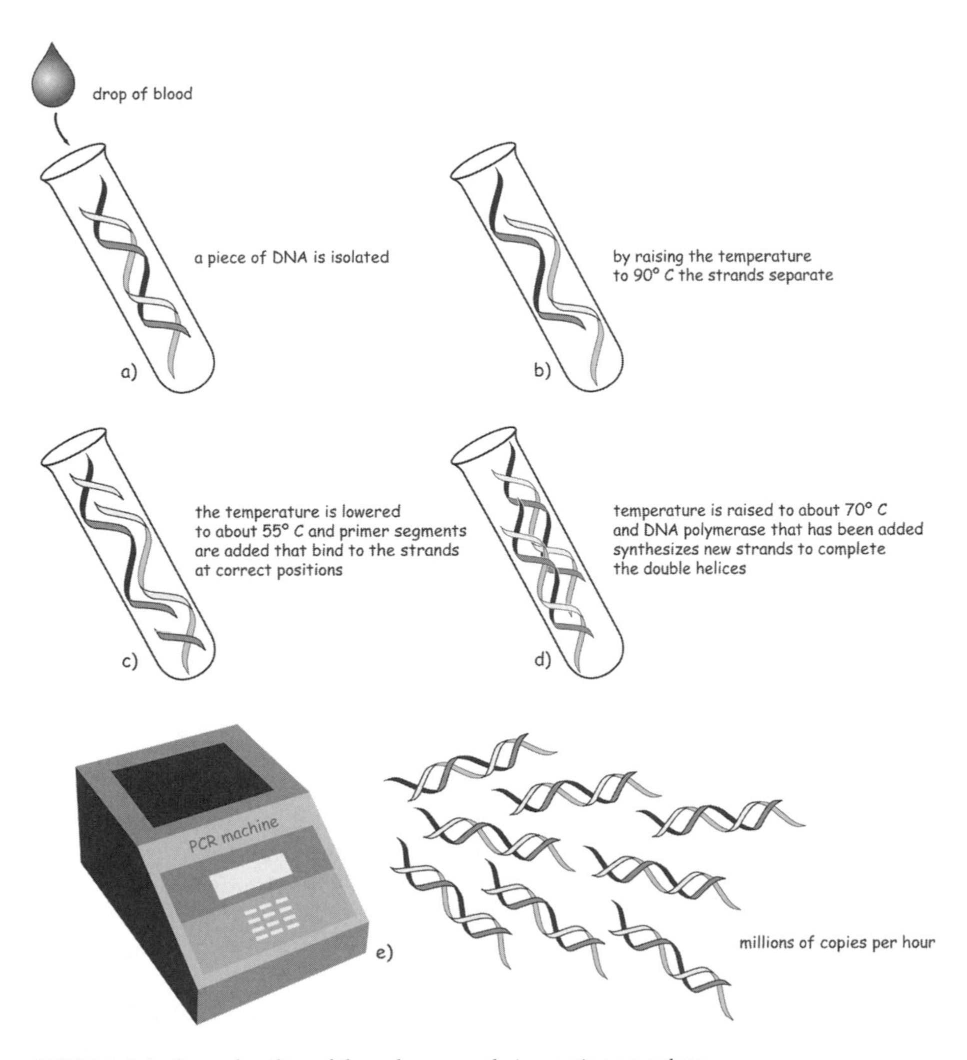

FIGURE 7.2 General outline of the polymerase chain reaction procedure.

Whole-Individual Cloning by Quasi-natural Mechanisms

Under this category of clonality, scientists promote or stimulate the production of whole-animal clones by mechanisms that are otherwise more or less natural. For almost every operation that nature has invented for generating whole-animal vertebrate clones, humans have found ways to prod things along by contrived manipulations or artificial stimuli. Examples are presented below. Some of these mechanisms fall into gray areas between what happens quite naturally, in one vertebrate species or another, and what is truly unparalleled in nature (to be discussed later).

Induced Parthenogenesis

In vertebrates, parthenogenesis (reproduction without the involvement of males or sperm) occurs constitutively in several taxa of squamate reptile, and sporadically in miscellaneous other species of reptiles, fish, and birds (see chapter 3). Natural parthenogenesis is unknown in mammals, perhaps because genomic imprinting, which seems to be unique to these animals, presents high genetic hurdles to unisexual reproduction (see box 2.2). Nevertheless, motivated in part by a desire to find alternative ways to generate embryonic stem cells (e.g., Trounson, 2002; K. Kim et al., 2007), scientists have discovered ways to overcome at least some of the genetic impediments to mammalian parthenogenesis (Rougier and Werb, 2001).

For instance, by altering the expression of loci that control genomic imprinting, Tomohiro Kono and colleagues (2004) were able to stimulate unfertilized, but nevertheless diploid, mouse oocytes to undergo parthenogenetic cell divisions. Similar reports for rabbits (Liu et al., 2002), cows (Morito et al., 2005), sheep (Alexander et al., 2006), pigs (Hao et al., 2006), camels (Mesbah et al., 2004), marmosets (Marshall et al., 1998), monkeys (Vrana et al., 2003), and even humans (Koh et al., 2004; Sengoku et al., 2004) show that various artificial stimuli, including electrical pulses (e.g., Ozil, 1990), chemical treatments (Mitalipov et al., 1999), and temperature shifts (Chang, 1954), can sometimes activate parthenogenetic cell divisions from mammalian oocytes (Kaufman, 1983). In most cases, only blastulas or early embryos were the final outcome of these experiments, but in some cases parthenogenetic offspring were born that could survive into adulthood (e.g., Kono et al., 2004).

The stage for such experimental inductions of parthenogenesis had been set more than a century earlier by the German-American scientist Jacques Loeb (1899). Working with otherwise sexual species of sea urchin, Loeb found that he could provoke embryonic development from unfertilized eggs by placing the animals in altered salt solutions. Interestingly, Loeb's research at the turn of the twentieth century was widely publicized by the media (Fangerau, 2005), and Loeb was sometimes criticized as "playing God" with life—much as genetic engineers at the turn of the twenty-first century receive public scrutiny and sometimes condemnation for their manipulations of life.

In some of the mammalian examples mentioned above, the diploid oocytes used in the experimental manipulations were of natural germ-line origin. In other cases, they carried nuclear genomes that researchers had physically injected from diploid donor cells (by nuclear transfer techniques to be described later). In both situations, the parthenogenetically activated cells underwent roughly normal development to the blastocyst stage or beyond. This illustrates how the distinctions can sometimes blur between "natural" and "artificial" processes used to generate whole-animal clones.

Induced Gynogenesis, Hybridogenesis, and Androgenesis

Gynogenetic and hybridogenetic lineages have arisen spontaneously following natural hybridization between various species of cold-blooded vertebrate (see chapter 4), so it is hardly surprising that researchers have tried to emulate these outcomes by forcing hybrid crosses in the laboratory. Indeed, hybridogenetic lineages of hemiclonal *Poeciliopsis* have been generated successfully in laboratory experiments (see chapter 4). Although similar attempts to generate self-sustaining unisexual lineages in most other taxa have been unsuccessful (but see Arai et al., 1993; Arai and Mukaino, 1997), production of first-generation gynogens from arranged hybrid crosses has been accomplished routinely, notably in carp species (Stanley, 1976; Cherfas et al., 1994) and trout and salmon (Johnson and Wright, 1986; Galbreath and Thorgaard, 1995; Dannewitz and Jansson, 1996; Galbreath et al., 1997).

For example, Jon Stanley and colleagues (1976) produced gynogenetic offspring in the white amur (*Ctenopharyngodon idella*) by exposing more than one million amur eggs to carp (*Cyprinus carpio*) milt that the researchers had irradiated with high doses of ultraviolet (UV) rays. Such radiation deactivates DNA but does not totally disable the sperm cells, and so it still triggered the development of many eggs, albeit without contributing paternal DNA to the resulting offspring. Most of the amur embryos were haploid and died early in development, but about 10,000 that were probably diploid (and appeared to carry only maternal DNA) survived to at least several months of age.

The experimental induction of transitory gynogenesis using impotent sperm traces back to the early 1900s (Hertwig, 1911, cited in Parsons and Thorgaard, 1984). Exposing sperm to harmful radiation (UV, gamma rays, or X-rays) has been the most popular method of degrading paternal DNA and thereby inducing gynogenesis in fish (Purdom, 1969; Stanley and Sneed, 1974; Cherfas, 1981; Naruse et al., 1985) and amphibians (Nace et al., 1970; Trottier and Armstrong, 1976; Tompkins, 1978). Furthermore, the approach is not confined to heterospecific (different-species) sperm; many examples entail the use of irradiated milt from conspecific males without any requirement for interspecific hybridization (e.g., Streisinger et al., 1981). Indeed, in referring specifically to fish, Stanley and colleagues speculated that "probably gynogenesis will take place in any species if sufficient numbers of eggs are treated with DNA-denatured sperm" (1976, p. 132).

The success rate in producing viable gynogenetic offspring can be enhanced by additional experimental manipulations. For instance, a temporary exposure of sperm-activated eggs to heat shock (Gillespie and Armstrong, 1981) or to hydrostatic pressure (Gillespie and Armstrong, 1979; Streisinger et al., 1981) often results in a suppression of cell division such that many eggs become diploid (with two sets of maternal DNA) and the resulting progeny survive much better than when haploid (Streisinger et al., 1981).

Remarkably, researchers also have generated vertebrate individuals that are androgenetic, meaning that they carry only paternal DNA (McKone and Halpern, 2003). The procedure involves the artificial inactivation of eggs, rather than sperm, usually by irradiation (Ellinger et al., 1975; Arai et al., 1979). The genetically impotent eggs are then exposed to nonirradiated sperm, and sometimes treated via temperature shock or hydrostatic pressure to reestablish diploidy (Parsons and Thorgaard, 1985; Komen et al., 1991). Viable androgenetic progeny have been produced by these artificial means in several species of fish (Purdom, 1969; Parsons and Thorgaard, 1984; Bercsenyi et al., 1998; Nam et al., 2000) and amphibians (Gillespie and Armstrong, 1980; Briedis and Elinson, 1982).

Although the kinds of laboratory manipulations described above typically do not yield self-sustaining lineages that are gynogenetic or androgenetic per se, they can yield, in effect, clonal lines in merely two generations (Komen et al., 1993; Sarder et al., 1999). Each first-generation gynogen or androgen that is diploid, for example, normally came from a single haploid gamete and thus is a "doubled haploid" (Young et al., 1998) that is isogenic and highly homozygous. A second generation of such chromosomal manipulation (Thorgaard, 1983) can produce lines of multiple individuals that for most practical purposes are clonal (e.g., Streisinger et al., 1981; Scheerer et al., 1986; Young et al., 1996; but see Buth et al., 1995). In zebrafish (*Danio rerio;* fig. 7.3), rainbow trout (*Oncorhynchus mykiss*), and a few other model experimental animals (Bongers et al., 1998), clonal lines derived from ancestors in which gynogenesis or androgenesis had been induced are used widely to construct linkage maps (Young et al., 1998; Nichols et al., 2003) and to dissect the genetic and environmental components of variation in quantitative phenotypes (Scheerer et al., 1991; Cheng and Moore, 1997; Robison et al., 1999; Brown and Thorgaard, 2002; Thorgaard et al., 2003), including particular behaviors (Iguchi et al., 2001; Lucas et al., 2004).

Induced Polyembryony

Fertility Drugs

Fertility treatments tend to increase human twinning rates often by about twofold. Are such treatments therefore a form of polyembryony induction? No, because the resulting offspring are typically polyzygotic rather than monozygotic,

FIGURE 7.3 The zebrafish (*Danio rerio*) is a popular fish for experimental research. In addition to the standard sexual forms of the species, cloned specimens have been engineered using gynogenetic treatments, and also using the techniques of nuclear transplantation.

and the treatments have no stimulatory effect on polyembryony per se. Indeed, monozygotic twinning in humans seems to occur at a universal rate of about one in 300 pregnancies irrespective of factors such as maternal age, family history, ethnicity, and fertility drugs that are known to influence rates of dizygotic twin pregnancies (see Reddy et al., 2005, and references therein). A lack of connection between fertility treatments and clonality makes sense because most fertility drugs, such as clomiphene citrate, act by stimulating ovulation (the release of mature eggs from an ovary). Sometimes two or more eggs are released and fertilized, leading to polyzygotic (but not clonemate) progeny.

Embryo Splitting

In the late 1800s and early 1900s, the German embryologist Hans Spemann conducted pioneering experiments on animal morphogenesis by micromanipulating the fertilized eggs and early embryos of amphibians. His discoveries on cellular mechanisms and processes underlying animal development earned Spemann the 1935 Nobel Prize in Medicine (Tagarelli et al., 2004). Spemann is also known as "the father of cloning" because some of his experiments resulted in polyembryos. By splitting salamander blastomeres or "embryos" at the few-cell stage (using noosed strands of his baby son's hair), Spemann generated monozygotic twins in these animals. The procedure is called embryo splitting. (In other experiments in which he tightened the noose only partially and split the cells incompletely, Spemann also generated his famously bizarre newt larvae with, for example, two heads but only one trunk and tail.)

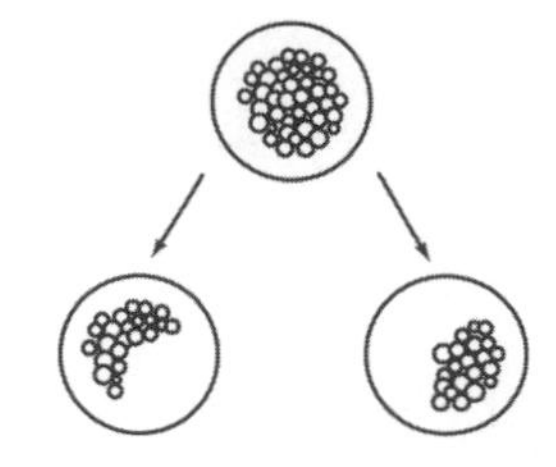

FIGURE 7.4 General outline of artificial polyembryony via embryo splitting.

In recent years, embryo splitting has been extended to experimental twin production (fig. 7.4) in mammals ranging from mice (Papaioannou et al., 1989) to cows (Seike et al., 1990) to rhesus monkeys (Chan et al., 2000). The procedure involves cleaving blastomeres (cells into which a zygote divides during early cleavage) into two or more subsets using microsurgical techniques, and transferring the clonemate cells into the wombs of surrogate mothers who sometimes carry the offspring to term. The usual rationale for such efforts is to produce genetically identical progeny either for scientific experimentation or for particular desired traits that the animals may possess. Reports also exist of human cloning—at least to the 32-cell stage of early development—by this procedure (Kolberg, 1993).

Forced Inbreeding

Before the rise of molecular biology in the middle and late twentieth century, the primary means by which humans generated "clones" was via multigeneration forced matings between close genetic relatives. Indeed, a critical feature of Gregor Mendel's work in the mid-1800s, which revealed the particulate nature of inheritance, is that the pea-plant strains that he crossed were partially inbred and "bred true" for the phenotypes of interest (tall versus short stature, round versus wrinkled seeds, etc.). Homozygosity within the starting lines greatly simplified genetic bookkeeping in the progeny of crosses between strains, and this enabled Mendel to deduce the basic rules of heredity without the complications that heterozygosity in the beginning lines would otherwise have entailed.

The utility of highly homozygous inbred strains for research on vertebrates has long been appreciated as well. In 1909, Clarence Little started the first inbred mouse strain (DBA) by mating brother and sister mice, and in the following decade several additional inbred strains (including C57BL and BALB/c) were initiated that are still employed widely today (Staats, 1966). In 1952, formal guidelines were first published for establishing inbred mouse strains: "A strain shall be regarded as inbred when it has been mated brother × sister for twenty or more consecutive generations. Parent × offspring matings may be substituted for brother × sister matings" (Carter et al., 1952, p. 603).

More than 450 inbred strains of house mouse (*Mus musculus*) are now available, each typically comprised of homozygous animals that in effect are isogenic clonemates (Beck et al., 2000). Many of these lines are used extensively in research, for example in crosses (analogous to those conducted by Mendel) to identify and map genes that underlie quantitative or qualitative traits (e.g., Darvasi, 1998), or as clonal replicates for assessing treatment effects in many biological experiments (Silver, 1995). Many of the inbred mouse lines were bred for phenotypes of special research interest, such as accelerated senescence, a preference for alcohol or narcotics, or genetic defects that lead to particular medical conditions such as premature retinal degeneration. Strains that are inbred to varying degrees have also been produced in various other animals, including cattle (Russell et al., 1984), sheep (Ercanbrack and Knight, 1993), and pigs (Zeng and Zeng, 2005).

Animal "cloning" via intense inbreeding is not without major shortcomings. First, inbred lines are difficult to establish in many species because of inbreeding depression (see box 6.5), and some strains that do survive may be quite debilitated. Second, heterosis (a higher fitness of heterozygous genotypes) can make it difficult to purge heterozygosity completely. Finally, the generations of forced mating required to establish a highly inbred line—20 or more—not only take much time and effort by animal breeders, depending on the generation length and husbandry of the target species, but can also entail an alteration or loss of traits of interest via genetic drift in the small captive population or via adaptation to the captive environment (Koide et al., 2000). It is for such reasons that whole-animal cloning via induced gynogenesis or androgenesis (a two-generation procedure, as previously described) has become, in suitable situations, a welcome alternative to traditional methods of creating highly inbred lines (Lucas et al., 2004).

Whole-Individual Cloning by Mechanisms Unknown in Nature

Nonhuman Animals

In the late 1990s, a British research team led by Ian Wilmut rocked the scientific world by announcing that they had cloned a lamb, called Dolly, from a somatic cell of an adult ewe (Wilmut et al., 1997). The procedure—which has no known analogue in nature—is outlined in figure 7.5, and it is straightforward in concept. Using tiny needles and micromanipulation techniques, the researchers removed the nucleus, with its entire DNA contents, from a mammary gland of an adult Finn Dorset ewe, transferred it into an enucleated egg cell from a Scottish Blackface, and stimulated the egg (with electrical impulses) to respond as if it had been fertilized. The egg began to develop into an early embryo that was returned to the womb of the Blackface ewe. Several months later, the ewe gave birth to the now-famous Dolly, a genetic clone of her Finn Dorset genetic dam. (Strictly, the clonal identity of Dolly to her mother applies only to her nuclear DNA; Dolly's mitochondrial DNA, housed in the cytoplasm, was inherited from the Blackface ewe.)

Similar nuclear transplantation (NT) experiments had been conducted decades earlier, specifically on amphibians. In the 1950s, embryologists Robert Briggs and Thomas King transferred cell nuclei from embryos or tadpoles of the leopard frog (*Rana pipiens;* fig. 7.6) into enucleated frog eggs and thereby constructed clonal embryos (some of which survived to the tadpole stage), each with a nuclear genome identical to that of its sole parent (Briggs and King, 1952; King and Briggs, 1955). In the 1970s, John Gurdon and colleagues extended the approach using nuclei artificially transplanted from skin cells of adult frogs (Gurdon et al., 1975). And in 1986, more than ten years before Dolly, Steen Willadsen (1986) used NT methods to create cloned sheep embryos using embryonic cells as nuclear donors. Over the years, similar kinds of success in cloning animals, at least to the blastocyst stage, were reported in the rabbit (Bromhall, 1975), mouse (Modlinsky, 1978; Illmensee and Hoppe, 1981), pig (Prather et al., 1989), rhesus monkey (Meng et al., 1997; Byrne et al., 2007), and several other vertebrate species (Galli et al., 2003b). Artificial cloning by nuclear transplantation thus had a long (half-century) pre-Dolly history (Gurdon and Byrne, 2003).

Despite such earlier cloning feats, the procreation of Dolly astonished almost everyone because biologists had assumed that the well-differentiated cells of

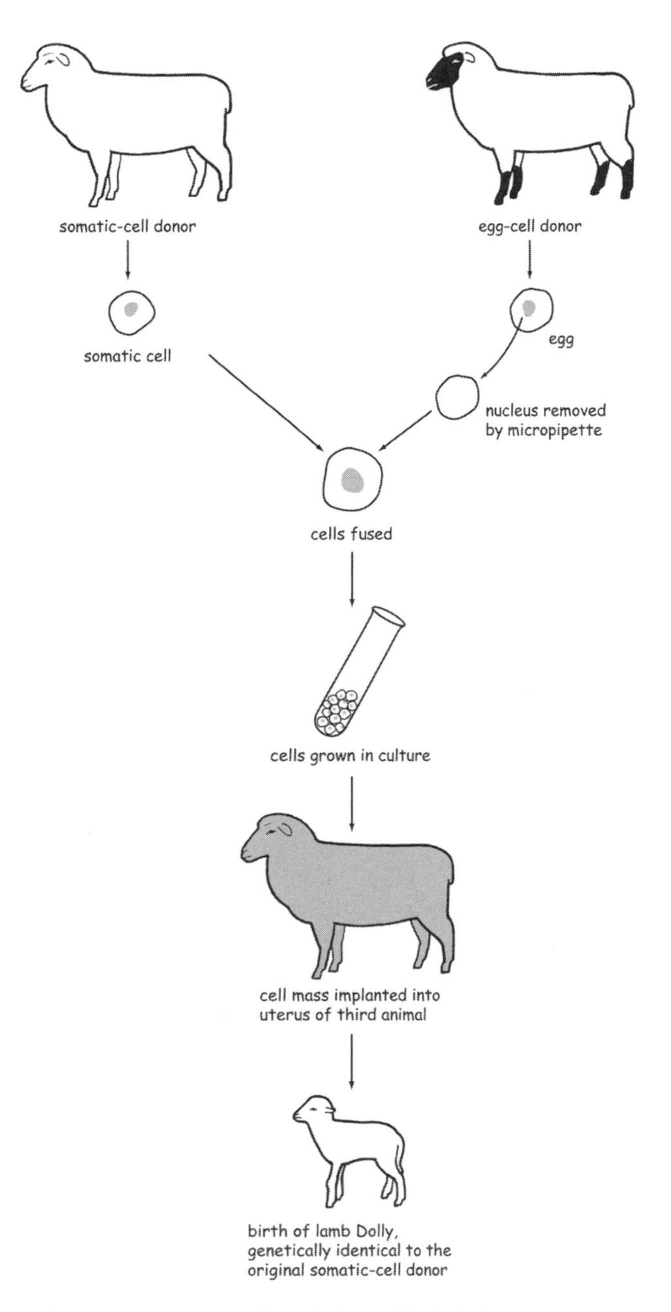

FIGURE 7.5 General outline of the artificial cloning procedure for the lamb Dolly.

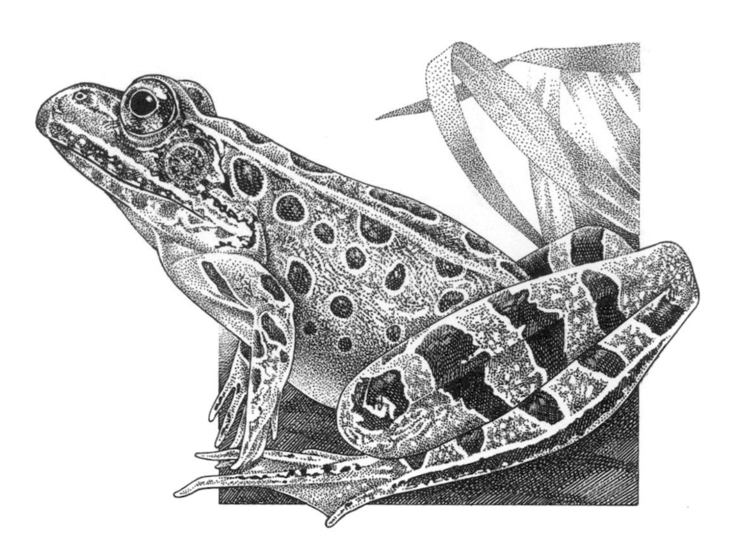

FIGURE 7.6 The leopard frog, *R. pipiens,* a species used in many of the original NT cloning methods.

adult mammals would have lost all capacity to supervise any process as complicated as embryogenesis. During the genetic and cellular symphonies of development in any multicellular organism, whole ensembles of genes are precisely activated, silenced, or otherwise modulated in tissue-specific fashion such that any one cell type—such as in a liver, kidney, or heart—expresses only a subset of its nuclear genome, and presumably lacks the wherewithal to orchestrate the complete ontogenetic process. In other words, the stodgy old genome of a differentiated cell, in contrast to the exuberant young genome of an embryonic stem cell, was presumed to have lost its former totipotent vigor (i.e., its capacity to diversify into all types of tissues and organs).

Dolly's production showed that at least some circumstances exist in which the genomes of differentiated adult cells can direct the development of a fully viable individual. In Dolly's case, the udder cells removed from her genetic dam first were grown in artificial culture under nutrient-poor conditions, forcing them to enter a quiescent state. In this resting stage of the cell cycle, most genes are inactive, and this seemed to be a key factor in the genome's ability, when placed into the refreshing intracellular environment of an egg, to reprogram its patterns of gene expression and thereby regain its youthful totipotency.

Just six months after Dolly was created, a Holstein calf named Gene was created through similar NT cloning techniques, and reports of several more cloned bovines quickly followed (Cibelli et al., 1998; Kato et al., 1998; Kubota et al., 2000). Like Dolly, these cloned animals often seemed to be healthy and normal (Lanza et al., 2001). Furthermore, some of the nuclei used in the bovine cloning process were themselves transgenic, having first been engineered by recombinant DNA procedures (as described earlier in the "Gene Cloning" section). A genetically modified cow named Annie was engineered as follows: First, cow cell nuclei were genetically altered to carry a bacterial transgene specifying lysostaphin, a protein that offers protection against *Staphylococcus* bacteria that in cattle can cause a mammary gland infection known as mastitis. The nucleus from one of these transgenic cells was transferred into an enucleated bovine egg, which was then implanted into the womb of Annie's surrogate mother. Annie was born a few months later, her cells genetically identical to the engineered construct that began the process. This example illustrates how whole-animal cloning piggybacked onto gene cloning might be used to generate herds of clonemate animals with desired genetic traits. Indeed, this is the commercial goal of many animal-cloning projects (Westhusin et al., 2001).

At the time of this writing, whole-animal clones have been produced by artificial nuclear transplantation in about 20 mammalian species (Cibelli, 2007; Wadman 2007). Besides the cow, these include the mouse (Wakayama et al., 1998), rat (Zhou et al., 2003), goat (Baguisi et al., 1999), pig (Betthauser et al., 2000; Polejaeva et al., 2000), wolf (M.K. Kim et al., 2007), dog (Lee et al., 2005), mouflon (Loi et al., 2001), African wildcat (Gómez et al., 2004), domestic cat (Shin et al., 2002), water buffalo (Kitiyanant et al., 2001), gaur (Lanz et al., 2000a), rabbit (Chesne et al., 2002), horse (Galli et al., 2003a), mule (Woods et al., 2003), and ferret (Li et al., 2006). The NT procedure has also been used successfully to clone other vertebrate animals, such as the zebrafish (Lee et al., 2002). More examples will likely appear before this book reaches print.

The diversity of animals (fig. 7.7) artificially cloned in the decade since Dolly might seem to imply that the NT procedure is simple and fail-safe, but this is not true. The techniques are complex and delicate at best. The domestic dog, among others, has proven exceedingly hard to clone for various technical reasons, and even for relatively "easy" species such as sheep, failed attempts (including deaths of NT eggs, blastocysts, and embryos) vastly outnumber individuals that survive to birth or beyond.

In some cases, artificial cloning has involved cross-species transplantation of nuclei, for example in the production of the water buffalo (*Bubalus bubalis*) by surrogate mother cows (Kitiyanant et al., 2001), the mouflon (*Ovis musimon*) by domestic sheep (Loi et al., 2001), and the African wildcat (*Felis silvestris*) by domestic cats (Gómez et al., 2004). Some scientists believe that NT cloning might play a useful role in last-ditch efforts to rescue critically endangered species from extinction (Lanza et al., 2000a, 2000b; Stone, 2006).

FIGURE 7.7 A menagerie of mammals artificially cloned using NT techniques.

Humans

In late 2002, two independent groups announced the birth—or the imminent birth, in one case—of the first human clones generated artificially by NT techniques. Just a few months later, one of these research teams (from the company Clonaid) reported another engineered human clone. Although these announcements attracted media attention and considerable public interest, the reports lacked scientific documentation and perhaps were hoaxes (Boese, 2003; Schatten et al., 2003). Whether or not the clones were bona fide, the fact remains that human cloning by NT methods is technically plausible, and it seems nearly inevitable that someone will someday accomplish the feat. For better or for worse, the production of human clones by artificial genetic or cellular manipulation will be a memorable milestone in the field of genetic engineering.

Any endeavor to generate cloned babies runs counter to strong public sentiment, and against the edicts of many agencies, governments, and some religions around the world. Many people view human cloning as ethically repugnant, going against nature's or God's design by undermining our deep-rooted values of individuality, uniqueness, and personal worth. Human cloning often has been portrayed as a means for rich or narcissistic individuals to make copies of themselves, or for unscrupulous dictators to produce brigades of identical soldiers. Most animal-cloning experts are also revolted by the prospects of human reproductive cloning, and they point out medical dangers associated with the process. As noted by Ian Wilmut, inventor of farm-animal cloning, 276 failures accompanied the production of Dolly the sheep, and washout rates at least this high can be expected in any human-cloning effort. Opponents of reproductive cloning contend that because the process leaves behind aborted zygotes, blastocysts, embryos, or fetuses, attempts at cloning are a moral indictment against humanity. Questions also arise about the quality of life and longevity of cloned people that escape such early sources of mortality.

How do proponents of human reproductive cloning defend their efforts? They contend that infertility is a medical disability, and that their goal is to help infertile people have children of their own. They suggest that the opportunity to reproduce is a basic human right, that cloning could become a genetic reproductive option for infertile people, and that society's ethical concerns are misplaced because a cloned baby is merely a "belated twin" of its predecessor—fundamentally no different from monozygotic twins that commonly arise during normal sexual reproduction. Proponents of human cloning also suggest that with diligent research, the technical hurdles can be overcome such that, in time, the procedures will become medically safe. Thus many advocates view human cloning as an assisted reproductive technology to be welcomed into the arsenal of medical ameliorations for common infertility problems.

"Therapeutic cloning" is another (and indeed the primary) rationale for cloning efforts in humans. Unlike "reproductive cloning," where the goal is to produce viable individuals, the idea underlying therapeutic cloning is to generate replacement tissues or organs for medical purposes. The procedure might work as follows: A suitable cell is taken from a patient, and its nucleus is inserted physically into an enucleated egg. The egg then begins to multiply in a test tube, and, from the developing cellular mass, pluripotent cells (i.e., stem cells that possess a capacity to differentiate into multiple tissue types) are recovered. Media reports of the first successful experimental trials of this form of human cloning appeared in early 2008, as this book went to press. The ultimate goal of therapeutic cloning might be, for example, to generate skin cells that could be used to repair damages from burns, pancreatic cells to treat diabetes, or retinal cells to reverse macular degeneration. When returned to the patient's body, the cloned cells would rejuvenate a damaged tissue or organ without evoking an immunological rejection, because the transferred cells would be clonally identical to those of the patient's body.

Reproductive cloning and therapeutic cloning are sometimes confused in debates about the ethics of artificial clonality in humans. The first laboratory step—inserting the nucleus of a somatic cell into an unfertilized egg—is the same in both cases, but otherwise the goals and approaches differ dramatically. In reproductive cloning, a genetically modified egg would be reimplanted in a womb and allowed to develop eventually into a full-blown and independent human being; in therapeutic cloning, an early clump of preimplantation cells that derives from a genetically modified egg would be grown in vitro and used to produce replacement cells or tissues for medical rehabilitation. Science will undoubtedly produce the technologies for both therapeutic and reproductive cloning in humans, but societies will have to decide the ethical merit or demerit of these activities.

SUMMARY OF PART IV

1. Scientists have co-opted and modified numerous natural genetic processes to generate clones under laboratory conditions. Some of these clonal methods apply at the levels of genes and particular pieces of DNA, whereas others apply to whole animals. Some of the laboratory procedures are close analogues of what happens in nature in one species or another, but others differ quite fundamentally from known routes to clonality in the wild.

2. Biotechnologists routinely clone short stretches of DNA, such as particular genes, using both in vivo and in vitro methods. The in vivo procedure involves cutting and pasting a gene of interest (e.g., from a vertebrate animal) into a bacterium or other biological vector that then copies the transgene to high numbers by normal processes of DNA replication and cell proliferation. The in vitro procedure involves replicating a gene to many copies in a test tube, using the polymerase chain reaction, or PCR.

3. Whole-animal clones have been produced in several vertebrate species using laboratory methods that induce parthenogenesis, gynogenesis, androgenesis, or hybridogenesis. The induction of parthenogenesis usually involves chemical treatments or other experimental manipulations that stimulate mitotic divisions in oocyte cells, without benefit of sperm. The artificial induction of sperm-dependent forms of clonality usually involves radiation treatments that inactivate

egg or sperm nuclei prior to fertilization; this is followed by exposure to high temperature or pressure, which triggers a diploid condition in the egg from which progeny arise that, in effect, have a single genetic parent.

4. In several vertebrate animals, including salamanders and various mammals, polyembryony has also been induced, experimentally, by microsurgical embryo splitting. In the mammalian cases, the separated cells are then implanted into the uteri of surrogate mothers for further embryonic development into clonemate progeny.

5. Another manipulated form of clonal propagation involves intense multigenerational inbreeding. When close relatives such as brothers and sisters are mated for more than 20 consecutive generations, highly homozygous strains arise whose members are, in essence, genetically identical to one another. Several such lines of inbred mice, for example, are used widely in genetic research.

6. Whole-animal clones can also be produced by nuclear transfer methods in which a diploid nucleus from a donor somatic cell is transplanted into a microsurgically enucleated egg, which then is artificially implanted into the womb of a surrogate mother. Such methods were introduced in the mid-twentieth century, but in the 1990s they received wide fanfare when Dolly the sheep was produced using a donor nucleus from a differentiated cell in an adult animal. Nuclear transfer cloning, sometimes involving a surrogate mother from a related taxon, has recently been accomplished in about 20 mammalian species.

EPILOGUE

Vertebrate clones are fascinating curiosities of nature, but they also offer scientific lessons about broader evolutionary processes. For example, most or all clonal lineages of extant vertebrates are evolutionarily young and have not adaptively radiated into multitaxon clades; this indicates that clonality in such animals is seldom a successful long-term strategy. Apart from lacking a strong genetic capacity to adapt to changing environments, clonal lineages probably deteriorate over time by the endogenous operation of Muller's ratchet. Clonality can be a highly successful tactic over the short term, though, as gauged by the local abundances and sometimes wide geographic distributions that some vertebrate clonal lineages have achieved. Immediate benefits to clonal reproduction—including the intact propagation of selection-tested genotypes and the avoidance of difficulties associated with sexuality—mean that vertebrate clonality can be selectively advantageous in particular ecological and genetic circumstances. This sentiment is bolstered by the many clonal vertebrate lineages of independent evolutionary origin that persist today. Another aspect of the capitalistic nature of vertebrate clonality is that all unisexual lineages appear to have arisen spontaneously, following hybridization between particular pairs of sexual species. Clonal vertebrate lineages thus provide a potent reminder that not all biological outcomes register slow and gradual evolutionary change.

Departures from strict clonality in vertebrates also offer evolutionary lessons. For instance, several and perhaps most gynogenetic and hybridogenetic lineages successfully incorporate sperm-derived DNA at least occasionally, thus making them less than strictly clonal or hemiclonal. Similarly, in the only known vertebrate that routinely displays clonal lineages by pathways of intense inbreeding (the mangrove killifish), occasional outcrossing presumably confers many of the evolutionary benefits of sexuality. And in the only vertebrate group that engages in constitutive polyembryony (*Dasypus* armadillos), sexual reproduction remains a major component of the life cycle. All such examples imply that elements of sexuality remain adaptively important to these otherwise clonal organisms.

In the modern age of genetic engineering, vertebrate clones are also generated routinely under human auspices. Some of the laboratory techniques are merely slight extensions of what can happen in nature; others are dramatic departures. The technical wherewithal to produce artificial clones in almost any vertebrate species, including *Homo sapiens,* is now or will soon be available. Science by itself cannot determine what is ethically proper, nor does naturalness (as opposed to artificiality) bear any necessary connection to moral propriety. Nevertheless, science can improve and contextualize our understandings of biology and nature. I hope this book has performed this role for many readers by removing much of the scientific mystique, but none of the awe, from the marvelous phenomenon of vertebrate clonality.

GLOSSARY

adaptation Any feature (e.g., morphological, physiological, behavioral) that helps an organism to survive and reproduce in a particular environment.

aging *See* senescence.

allele Any of the possible forms, or classes of forms, of a specified gene. A diploid individual carries two alleles at each autosomal gene, and these can either be identical in state (in which case the individual is homozygous) or different in state (heterozygous). At each autosomal gene, a population of N diploid individuals harbors $2N$ alleles, various of which may differ in details of nucleotide sequence.

allopatric Inhabiting different geographic areas.

ameiotic *See* apomixis.

amino acid Any of about 20 different organic molecules found universally in proteins, and displaying carboxyl and amino groups, that when joined together form a polypeptide.

amphimixis A sexual life cycle (characteristic of most multicellular species) with alternating syngamy and meiosis.

androdioecy A condition in which hermaphrodites and males coexist within a species.

androgenesis Reproduction in which offspring—typically diploid—carry nuclear DNA only from the male parent.

apomixis Asexual reproduction without meiosis or fertilization.

asexual reproduction Any form of reproduction that does not involve the fusion of sex cells (gametes).

automixis A form of "asexual" reproduction that entails the union of meiotic products of an individual. (Note: Some authors use the term more broadly to encompass any form of uni-individual reproduction that includes meiosis or a meiosis-type process, including premeiotic endomitosis; see Haccou and Schneider, 2004.)

autosome A chromosome in the nucleus other than a sex chromosome. In diploid organisms, autosomes are present in homologous pairs.

backcross *n.*, the progeny from a cross between an offspring and one of its parents; *v.*, to conduct such a mating.

bacteriophage A virus that infects a bacterium.

bacterium (pl. bacteria) A unicellular microorganism without a true cellular nucleus.

biodiversity Life's genetic heterogeneity, at any or all levels of biological organization.

biotechnology The use of living entities or their components or products for industrial or commercial applications.

biotype A group of unisexual or clonal organisms, for which the concept of biological "species" does not readily apply.

bisexual A population or species composed of male and female individuals.

blastocyst A mammalian embryo before its implantation into the uterine wall.

blastomere One of the cells into which an egg divides during cleavage.

blastula A hollow sphere of cells resulting from early cell cleavages from a zygote.

bottleneck A severe but temporary reduction in population size.

cell A small, membrane-bound unit of life that is usually capable of self-reproduction.

chorion A fluid filled sac in which vertebrate embryos develop.

chromosome A threadlike structure within a cell that carries genes.

classification (biological) A process of establishing, defining, and ranking biological taxa within hierarchical groups; alternately, the outcome itself of this process.

clone *n.*, a biological entity (e.g., gene, cell, or multicellular organism) that is genetically identical to another; alternately, all genetically identical entities that have descended asexually from a given ancestral entity; *v.*, to produce such genetically identical entities or lineages.

clonemates Two or more organisms that are genetically identical.

coevolution The interdependent evolution of two or more interacting species or genomes.

congeneric Belonging to the same taxonomic genus.

conspecific Belonging to the same taxonomic species.

constitutive Of consistent and essential occurrence. *See also* facultative.

cytonuclear analysis A genetic appraisal based on information jointly from an organism's cytoplasmic genome (typically mtDNA) and one or more loci in the nuclear genome.

cytoplasm The portion of a cell outside the nucleus.

deoxyribonucleic acid (DNA) The genetic material of most life-forms; a double-stranded molecule composed of strings of nucleotides.

diapause A period of inactivity and suspension of growth.

dioecy A condition in which males and females are separate individuals (often used in the botanical literature).

diploid A usual somatic cell condition wherein two copies of each chromosome are present.

dizygotic twins (or fraternal twins) Genetically nonidentical siblings that stem from two separate zygotes during a pregnancy.

ecology The study of the interrelationships among living organisms and their environments.

ectoparasite A parasite that remains on the outside of a host's body.

egg A female gamete. *See also* oocyte (unfertilized).

electrophoresis The movement of charged proteins or nucleic acids through a supporting gel under the influence of an electric current.

embryo An organism in the early stages of development (in humans, usually up to the beginning of the fourth month of pregnancy).

embryogenesis The development of an embryo.

embryonic stem cell *See* stem cell.

endangered species A species at immediate risk of extinction.

endemic Native to, and restricted to, a particular geographic area.

endogenous Produced or naturally occurring within the body.

endomitosis Chromosomal replication within a cell that does not divide.

endoparasite A parasite that resides inside a host's body for at least part of its life cycle.

enzyme A catalyst (normally a protein) of a specific chemical reaction.

eukaryote Any organism in which chromosomes are housed in a membrane-bound nucleus.

evolution Change across time in the genetic composition of a population or species.

exogenous Produced or naturally occurring outside the body.

exon A coding segment of a gene. *See also* intron.

expression (of a gene) Activation of a gene to begin the process (RNA formation) that later may eventuate in production of a protein.

extant Alive today.

extinction The permanent disappearance of a population or species.

facultative Optional; occurring only part of the time. *See also* constitutive.

fertilization The union of two gametes to produce a zygote.

fetus An individual at intermediate stages of development in the uterus (in humans, beginning at about the fourth month of pregnancy).

fitness (Darwinian) The contribution of an individual or a genotype to the next generation relative to the contributions of other individuals or genotypes. *See also* inclusive fitness.

fraternal twins *See* dizygotic twins

gamete A mature reproductive sex cell (egg or sperm).

gametogenesis The process by which sex cells are produced.

gene The basic unit of heredity; usually taken to imply a sequence of nucleotides specifying production of a polypeptide or other functional product, but can also be applied to stretches of DNA with unknown or unspecified function.

genealogy A record of descent from ancestors through a pedigree.

gene flow The geographic movement of genes, normally among populations within a species.

gene pool The sum total of all hereditary material in a population or species.

genetic Of or pertaining to the study of heredity.

genetically modified organism (GMO) Any creature whose genes have been deliberately and directly altered by humans.

genetic drift Change in allele frequency in a finite population by chance sampling of gametes between generations.

genetic engineering The direct and purposeful alteration of genetic material by humans.

genetic load The burden to a population of deleterious genes.

genetic markers Natural nucleic-acid or protein tags that exist in all forms of life.

genome The complete genetic constitution of an organism; can also refer to a particular composite piece of DNA such as the mitochondrial genome.

genotype The genetic constitution of an individual with reference to a gene or set of genes.

germline The lineage of cells leading to an individual's gametes.

gonochoristic A sexual system in which each individual is either a male or a female.

group selection Natural selection acting upon groups of individuals via differences in the traits that those groups possess.

gynodioecy A condition in which hermaphrodites and females both occur within a species.

gynogen An individual or strain that reproduces by gynogenesis.

gynogenesis Reproduction in which a sperm cell is needed to activate cell divisions in an oocyte but the resulting offspring carry nuclear DNA only from the female parent.

haploid The usual condition of a gametic cell in which one copy of each chromosome is present.

hemiclone The portion of a genome that is transmitted intact, without recombination, in a hybridogenetic lineage.

heredity The phenomenon of familial transmission of genetic material from one generation to the next.

hermaphrodite An individual that produces both male and female gametes, either sequentially or simultaneously.

heterogametic sex The sex that produces gametes that each contain one of two different types of sex chromosomes.

heterosis Higher genetic fitness of heterozygotes than homozygotes.

heterospecific Of or pertaining to another species.

heterozygosity The percentage of heterozygotes, or the percentage of loci in heterozygous state, in a population.

heterozygote A diploid organism possessing two different alleles at a specified genetic locus.

homogametic sex The sex that produces gametes that all contain the same type of sex chromosome.

homology Similarity of traits (e.g., morphological, molecular, etc.) due to inheritance from a shared ancestor.

homospecific Of or pertaining to the same species.

homozygote A diploid organism possessing two identical alleles at a specified genetic locus.

hybridization The successful mating of individuals belonging to genetically different populations or species.

hybridogen An individual or strain that reproduces by hybridogenesis.

hybridogenesis A quasi-sexual form of reproduction in which egg and sperm fuse to initiate embryonic development, but germ cells in the offspring later undergo an abnormal meiosis in which the resulting gametes carry no paternally derived genes.

implantation The attachment of an embryo to the uterine wall.

inbreeding The mating of kin.

inbreeding depression A loss in genetic fitness due to inbreeding.

inclusive fitness An individual's own genetic fitness as well as his or her effects on the genetic fitness of close relatives.

insecticide A pesticide applied to insects.

introgression The movement of genes between species via hybridization.

intron A noncoding portion of a gene. *See also* exon.

invertebrate An animal that does not possess a backbone.

in vitro Outside the living body—for example, in a laboratory or test tube.

in vivo Within a living body.

kin selection A form of natural selection due to individuals favoring the survival and reproduction of genetic relatives.

kleptogenesis Reproduction by gynogenesis-type or hybridogenesis-type mechanisms but with at least occasional incorporation of sperm-derived DNA into the otherwise clonal lineage.

life cycle The sequence of events for an individual, from its origin as a zygote to its death; one generation.

ligase An enzyme that catalyzes the binding together of DNA pieces end to end.

linked genes Loci carried on the same chromosome.

locus (pl. loci) A gene; a location on a chromosome.

matriline A genetic transmission pathway strictly through females.

meiosis The cellular process whereby a diploid cell divides to form haploid gametes.

meiotic Of or pertaining to meiosis. *See also* automixis.

metabolism The sum of all physical and chemical processes by which living matter is produced and maintained, and by which cellular energy is made available to an organism.

microbe A very small organism visible only under a microscope.

mitochondrial DNA (mtDNA) Genetic material housed within the mitochondrion.

mitochondrion (pl. mitochondria) An organelle in the cytoplasm of animal and plant cells that is the site of some key metabolic pathways involved in cellular energy production.

mitosis A process of cell division that produces daughter cells with the same chromosomal constitution as the parent cell.

mixed mating A mating system that involves both selfing and outcrossing.

molecular clock An evolutionary timepiece based on the evidence that genes or proteins tend to accumulate mutational differences at roughly constant rates in particular lineages.

molecular marker *See* genetic markers.

monoecy A situation in which an individual plant has male and female reproductive organs and produces both pollen and ova.

monophyletic Of single evolutionary origin.

monozygotic twins Genetically identical siblings (barring mutation) that stem from a single zygote during a pregnancy. *See also* polyembryony.

morphology The visible structures of organisms.

mosaic An individual that carries two or more sets of genetically different cells.

Muller's ratchet The evolutionary accumulation of harmful mutations within a clonal lineage.

multicellular Composed of two or more cells.

mutation A change in the genetic constitution of an organism.

myoblast A muscle-producing cell.

natural selection The differential contribution by individuals of different genotypes to the next generation.

nepotism Favoritism directed toward kin.

neuron A nerve cell.

niche The ecological "role" of a species in a natural community; an organism's way of making a living.

nuclear transfer (NT) cloning The construction of genetically identical organisms by an artificial process that begins with the transfer of the nucleus of a somatic cell into an enucleated egg.

nucleic acid *See* deoxyribonucleic acid (DNA) *and* ribonucleic acid (RNA).

nucleotide A unit of DNA or RNA consisting of a nitrogenous base, a pentose sugar, and a phosphate group.

nucleus (pl. nuclei) A portion of a cell bounded by a membrane and containing chromosomes.

ontogeny The course of development and growth of an individual to maturity.

oocyte (unfertilized) A female gamete, also known as an egg cell, or ovum.

oogenesis The production of oocytes.

organ A part of an animal, such as the heart, that forms a structural and functional unit.

organelle A complex, recognizable structure in the cell cytoplasm (such as a mitochondrion or chloroplast).

outcrossing Mating with another, typically unrelated, individual.

oviparous Egg-laying.

ovum Egg.

paraphyly A situation in which an assemblage of organisms includes a common ancestor and some, but not all, of its evolutionary descendants.

parasite An organism that for at least some part of its life cycle is intimately associated with and harmful to a host. *See also* sexual parasitism.

parasitoid A parasite that usually feeds for part of its life cycle within a host's body, and that does not have multiple generations per host generation.

parthenogen An individual or strain that reproduces by parthenogenesis.

parthenogenesis The development of an individual from an egg without fertilization.

paternal leakage (of mtDNA) The occasional incorporation of sperm mtDNA into a zygote, and thereby into the resulting offspring.

pathogen An organism or microorganism that produces a disease.

pedigree A diagram displaying mating partners and their offspring across generations.

pesticide A chemical agent that kills animal pests.

phage *See* bacteriophage.

phylogenetic Of or pertaining to phylogeny.

phylogeny Evolutionary relationships (historical descent) of a group of organisms or species.

plasmid A small, extrachromosomal genetic element found in bacteria.

pleiotropy A phenomenon in which a single gene contributes to more than one distinct phenotype.

pollen A male gamete in plants.

polyembryony The production of genetically identical offspring within a clutch or litter.

polymerase An enzyme that catalyzes the formation of nucleic acid molecules.

polymerase chain reaction (PCR) A laboratory procedure for the in vitro replication of DNA.

polymorphism The presence of two or more genetically distinct forms (traits or genotypes) in a population.

polypeptide A string of amino acids.

polyphyletic A group of organisms perhaps classified together but traceable to different ancestors.

polyploidy A condition in which more than two sets of chromosomes are present within a cell.

population All individuals of a species normally inhabiting a defined area.

predator An organism that feeds by preying on other organisms.

premeiotic endomitosis *See* endomitosis.

primer (for PCR) A short string of nucleotides used in conjunction with an appropriate enzyme to initiate synthesis of a nucleic acid.

prokaryote Any microorganism that lacks a chromosome-containing, membrane-bound nucleus.

protandry A type of hermaphroditism in which an individual is first male and then later in life switches to female.

protein A macromolecule composed of one or more polypeptide chains.

protogyny A type of hermaphroditism in which an individual is first female and then later in life switches to male.

pseudogamy The clonal development of an ovum following stimulation (but not fertilization) by a male gamete.

recombinant DNA A new hereditary molecule that has arisen from genetic recombination.

recombination (genetic) The formation of new combinations of genes, as for example occurs naturally via meiosis and fertilization.

regulatory gene A segment of DNA that exerts operational control over the expression of other genes.

reproductive cloning The construction of genetically identical organisms via biotechnology.

restriction enzyme Any organic compound produced by a bacterium that catalyzes the cleavage of DNA molecules at specific recognition sites.

restriction fragment A linear segment of DNA resulting from cleavage of a longer segment by a restriction enzyme.

ribonucleic acid (RNA) The genetic material of many viruses, similar in structure to DNA. Also, any of a class of molecules that normally arise in cells from the transcription of DNA.

self-fertilization (selfing) The union of male and female gametes from the same hermaphroditic individual.

senescence A persistent decline with age in the survival probability or reproductive output of an individual due to interior physiological deterioration.

sex chromosome A chromosome in the cell nucleus involved in distinguishing the two sexes.

sexual parasitism The utilization by gynogenetic or hybridogenetic females of sperm from the males of other species.

sexual reproduction Organismal procreation via the generation and fusion of gametes.

somatic Of or pertaining to any cell (or body part) in a multicellular organism other than those destined to become gametes.

species (biological) Groups of actually or potentially interbreeding individuals that are reproductively isolated from other such groups.

sperm A male gamete in animals.

spermatogenesis The production of sperm.

stem cell An undifferentiated, mitotically active cell that serves to produce new cells or replenish those lost during the life of an individual. Those found in mature organisms are adult stem cells; those found in early life stages are embryonic stem cells.

sympatric Inhabiting the same geographic area.

synapsis The alignment of homologous pairs of chromosome during meiosis.

syngamy The genetic union of a male gamete and a female gamete.

systematics The comparative study and classification of organisms, particularly with regard to their phylogenetic relationships.

taxon (pl. taxa) A biotic lineage or entity deemed sufficiently distinct from other such lineages as to be worthy of a formal taxonomic name.

taxonomy The practice of naming and classifying organisms.

tetraploid Possessing four complete sets of chromosomes.

tissue A population of cells of the same type performing the same function.

totipotent Pertaining to cells capable of generating an entire organism.

transcription The cellular process by which an RNA molecule is formed from a DNA template.

transformation The introduction of foreign DNA into a cell or organism.

transgene Foreign DNA carried by a genetically modified organism.

transgenic organism A genetically engineered organism containing foreign DNA.

translation The cellular process by which a polypeptide chain is formed from an RNA template.

triploid Possessing three complete sets of chromosomes.

tychoparthenogenesis Sporadic or facultative parthenogenesis.

unisexual species A species consisting exclusively of females.

uterus The mammalian womb.

vertebrate An animal that possesses a backbone.

virus A tiny, obligate intracellular parasite, incapable of autonomous replication, that utilizes the host cell's replicative machinery.

viviparous Producing live offspring by giving birth from within the body of a parent.

W chromosome In birds, the sex chromosome normally present in females only.

X chromosome The sex chromosome normally present as two copies in female mammals (the homogametic sex), but as only one copy in males (the heterogametic sex).

Y chromosome In mammals, the sex chromosome normally present in males only.

Z chromosome The sex chromosome normally present as two copies in male birds (the homogametic sex), but as only one copy in females (the heterogametic sex).

zygote Fertilized egg; the diploid cell arising from the union of male and female haploid gametes.

REFERENCES

Abel, D.C., C.C. Koenig, and W.P. Davis. 1987. Emersion in the mangrove forest fish *Rivulus marmoratus,* a response to hydrogen sulfide. *Environ. Biol. Fishes* 18:67–72.

Abramoff, P., R.M. Darnell, and J.S. Balsano. 1968. Electrophoretic demonstration of the hybrid origin of the gynogenetic teleost *Poecilia formosa. Am. Nat.* 102:555–558.

Abuhteba, R.M., J.M. Walker, and J.E. Cordes. 2000. Genetic homogeneity based on skin histocompatibility and the evolution and systematics of parthenogenetic *Cnemidophorus laredoensis* (Sauria: Teiidae). *Can. J. Zool.* 78:895–904.

Adams, M., R. Foster, M.N. Hutchinson, R.G. Hutchinson, and S.C. Donnellan. 2003. The Australian scincid lizard *Menetia greyii:* a new instance of widespread vertebrate parthenogenesis. *Evolution* 57:2619–2627.

Alexander, B., G. Coppola, D. Di Berardino, G.J. Rho, E. St. John, D.H. Betts, and W.A. King. 2006. The effect of 6-dimethylaminopurine (6-DMAP) and cycloheximide (CHX) on the development and chromosomal complement of sheep parthenogenetic and nuclear transfer embryos. *Mol. Reprod. Dev.* 73:20–30.

Allard, R.W. 1975. The mating system and microevolution. *Genetics* 79:115–126.

Alves, M.J., M.M. Coelho, and M.J. Collares-Pereira. 1998. Diversity in the reproductive modes of females of the *Rutilus alburnoides* complex (Teleostei, Cyprinidae): a way to avoid the genetic constraints of uniparentalism. *Mol. Biol. Evol.* 15:1233–1242.

Alves, M.J., M.M. Coelho, and M.J. Collares-Pereira. 2001. Evolution in action through hybridization and polyploidy in an Iberian freshwater fish: a genetic review. *Genetica* 111:375–385.

Alves, M.J., M.M. Coelho, M.J. Collares-Pereira, and T.W. Dowling. 1997. Maternal ancestry of the *Rutilus alburnoides* complex (Teleostei, Cyprinidae) as determined by analysis of cytochrome *b* sequences. *Evolution* 51:1584–1592.

Alves, M.J., M.M. Coelho, M.I. Próspero, and M.J. Collares-Pereira. 1999. Production of fertile unreduced sperm by hybrid males of the *Rutilus alburnoides* complex (Teleostei, Cyprinidae): an alternative route to genome tetraploidization in unisexuals. *Genetics* 151:277–283.

Alves, M.J., M.J. Collares-Pereira, T.E. Dowling, and M.M. Coelho. 2002. The genetics of maintenance of an all-male lineage in the *Squalius alburnoides* complex. *J. Fish Biol.* 60:649–662.

Alves, M.J., M. Gromicho, M.J. Collares-Pereira, E. Crespo-López, and M.M. Coelho. 2004. Simultaneous production of triploid and haploid eggs by triploid *Squalius alburnoides* (Teleostei: Cyprinidae). *J. Exp. Zool.* 301:552–558.

Angers, B., and I.J. Schlosser. 2007. The origin of *Phoxinus eos-neogaeus* unisexual hybrids. *Mol. Ecol.* 16:4562–4571.

Angus, R.A., and R.J. Schultz. 1979. Clonal diversity in the unisexual fish *Poeciliopsis monacha-lucida:* a tissue graft analysis. *Evolution* 33:27–40.

Arai, K., K. Matsubara, and R. Suzuki. 1993. Production of polyploids and viable gynogens using spontaneously occurring loach, *Misgurnus anguillicaudatus. Aquaculture* 117:227–235.

Arai, K., and M. Mukaino. 1997. Clonal nature of gynogenetically induced progeny of triploid (diploid X tetraploid) loach *Misgurnus anguillicaudatus* (Pisces: Cobitidae). *J. Exp. Zool.* 278:412–421.

Arai, K., H. Onozato, and F. Yamazaki. 1979. Artificial androgenesis induced with gamma irradiation in masu salmon, *Oncorhynchus masou. Bull Fac. Fish. Hokkaido Univ.* 30:181–186.

Ashworth, C.J., A.W. Ross, and P. Barrett. 1998. The use of DNA fingerprinting to assess monozygotic twinning in Mieshan and Landrace x large white pigs. *Reprod. Fertil. Dev.* 10:487–490.

Atz, J.W. 1965. Hermaphroditic fish. *Science* 150:789–797.

Atz, J.W., and S. Kazianis. 2001. In appreciation of Klaus D. Kallman. *Mar. Biotechnol.* 3:S3–S5.

Avila, L.J., and R.A. Martori. 1991. A unisexual species of *Teius* Merrem 1820 (Sauria Teiidae) from central Argentina. *Trop. Zool.* 4:193–201.

Avise, J.C. 1989. Nature's family archives. *Nat. Hist.* 98:24–27.

Avise, J.C. 1991. Martriarchal liberation. *Nature* 352:192.

Avise, J.C. 1998. *The Genetic Gods: Evolution and Belief in Human Affairs.* Harvard University Press, Cambridge, MA.

Avise, J.C. 2004a. *Molecular Markers, Natural History, and Evolution.* 2nd ed. Sinauer, Sunderland, MA.

Avise, J.C. 2004b. *The Hope, Hype, and Reality of Genetic Engineering: Remarkable Stories from Agriculture, Industry, Medicine, and the Environment.* Oxford University Press, New York.

Avise, J.C. 2006. *Evolutionary Pathways in Nature: A Phylogenetic Approach.* Cambridge University Press, New York.

Avise, J.C., J.M. Quattro, and R.C. Vrijenhoek. 1992. Molecular clones within organismal clones: mitochondrial DNA phylogenies and the evolutionary histories of unisexual vertebrates. *Evol. Biol.* 26:225–246.

Avise, J.C., J.C. Trexler, J. Travis, and W.S. Nelson. 1991. *Poecilia mexicana* is the recent female parent of the unisexual fish *P. formosa. Evolution* 45:1530–1533.
Avise, J.C., and R.C. Vrijenhoek. 1987. Mode of inheritance and variation of mitochondrial DNA in hybridogenetic fishes of the genus *Poeciliopsis. Mol. Biol. Evol.* 4:514–525.
Avise, J.C., D. Walker, and G.C. Johns. 1998. Speciation durations and Pleistocene effects on vertebrate phylogeography. *Proc. R. Soc. Lond. B Biol. Sci.* 265:1707–1712.
Baer, J.G., and L. Euzet. 1961. Classe des monogenes. Pp. 241–325 in: *Traité de Zoologie, IV,* P. Grassé (ed.). Masson, Paris.
Bagatto, B., D.A. Crossley II, and W.W. Burggren. 2000. Physiological variability in neonatal armadillo quadrulets: within- and between-litter differences. *J. Exp. Biol.* 203:1733–1740.
Baguisi, A., E. Behboodi, D.T. Melican, J.S. Pollock, M.M. Destrempes, C. Cammuso, J.L. Williams, S.D. Nims, C.A. Porter, P. Midura, et al. 1999. Production of goats by somatic cell nuclear transfer. *Nat. Biotechnol.* 17:456–461.
Baker, H.C. 1948. Dimorphism and monomorphism in the Plumbagionaceae. I. Survey of the family. *Ann. Bot.* 12:207–219.
Baker, H.G. 1953. Dimorphism and monomorphism in the Plumbagionaceae. III. Correlation of geographical distribution patterns with dimorphism and monomorphism in *Limonium. Ann. Bot.* 17:615–627.
Baker, H.G. 1955. Self-compatibility and establishment after "long-distance" dispersal. *Evolution* 9:347–349.
Baker, H.G. 1965. Characteristics and modes of origin of weeds. Pp. 147–172 in: *Genetics of Colonizing Species,* H.G. Baker and G.L. Stebbins (eds.). Academic Press, New York.
Balsano, J.S., R.M. Darnell, and P. Abramoff. 1972. Electrophoretic evidence of triploidy associated with populations of the gynogenetic teleost *Poecilia formosa. Copeia* 1972:292–297.
Beck, J.A., S. Lloyd, M. Hafezpatast, M. Lennon-Pierce, J.T. Eppig, M.F.W. Festing, and E.M.C. Fisher. 2000. Genealogies of mouse inbred strains. *Nat. Genet.* 24:23–25.
Bell, G. 1982. *The Masterpiece of Nature: The Evolution and Genetics of Sexuality.* University of California Press, Berkeley.
Bercsenyi, M., I. Magyari, B. Urbanyi, L. Orban, and L. Horvath. 1998. Hatching out goldfish from common carp eggs: interspecific androgenesis between two cyprinid species. *Genome* 41:573–579.
Berger, A.J. 1953. Three cases of twin embryos in passerine birds. *Condor* 55:273–274.
Bernstein, C., and H. Bernstein. 1991. *Aging, Sex, and DNA Repair.* Academic Press, New York.
Betthauser, J., E. Forsberg, M. Augenstein, L. Childs, K. Eilertsen, J. Enos, T. Forsythe, P. Golueke, G. Jurgella, R. Koppang, et al. 2000. Production of cloned pigs from *in vitro* systems. *Nat. Biotechnol.* 18:1055–1059.

Beukeboom, L.W., and R.C. Vrijenhoek. 1998. Evolutionary genetics and ecology of sperm-dependent parthenogenesis. *J. Evol. Biol.* 11:755–782.

Beukeboom, L.W., R.P. Weinzierl, K.M. Reed, and N.K. Michiels. 1995. Amazon molly and Muller's ratchet. *Nature* 375:111–112.

Bi, K., and J.P. Bogart. 2006. Identification of intergenomic recombinations in unisexual salamanders of the genus *Ambystoma* by genomic *in situ* hybridization (GISH). *Cytogenet. Genome Res.* 112:307–312.

Bi, K., J.P. Bogart, and J. Fu. 2007a. Intergenomic translocations in unisexual salamanders of the genus *Ambystoma* (Amphibia, Caudata). *Cytogenet. Genome Res.* 116:289–297.

Bi, K., J.P. Bogart, and J. Fu. 2007b. Two rare aneutriploids in the unisexual *Ambystoma* (Amphibia, Caudata) identified by GISH indicating two different types of meiotic errors. *Cytogenet. Genome Res.* 119:127–130.

Bigelow, H.B. 1909. The Medusae. *Mem. Mus. Comp. Zool.* 37:9–245.

Billingham, R.E. 1959. Reactions of grafts against their hosts. *Science* 130:947–953.

Billingham, R.E., and W.B. Neaves. 2005. Exchange of skin grafts among monozygotic quadruplets in armadillos. *J. Exp. Zool.* 213:257–260.

Binet, M.-C., and B. Angers. 2005. Genetic identification of members of the *Phoxinus eos-neogaeus* hybrid complex. *J. Fish Biol.* 67:1169–1177.

Bird, A. 2007. Perceptions of epigenetics. *Nature* 447:396–398.

Birkhead, T.R., and A.P. Møller. 1993. Female control of paternity. *Trends Ecol. Evol.* 8:100–104.

Birky, C.W., Jr. 2001. The inheritance of genes in mitochondria and chloroplasts: laws, mechanisms, and models. *Annu. Rev. Genet.* 35:125–148.

Boese, A. 2003. *The Museum of Hoaxes: A History of Outrageous Pranks and Deceptions.* Plume, New York.

Bogart, J.P. 1989. A mechanism for interspecific gene exchange via all-female salamander hybrids. Pp. 170–179 in: *Evolution and Ecology of Unisexual Vertebrates,* R.M. Dawley and J.P. Bogart (eds.). New York State Museum, Albany.

Bogart, J.P. 2003. Genetics and systematics of hybrid species. Pp. 109–134 in: *Reproductive Biology and Phylogeny of Urodela,* D.M. Sever (ed.). M/S Science, Enfield, NH.

Bogart, J.P., K. Bi, J. Fu, D.W.A. Noble, and J. Niedzwiecki. 2007. Unisexual salamanders (genus *Ambystoma*) present a new reproductive mode for eukaryotes. *Genome* 50:119–136.

Bogart, J.P., R.P. Elinson, and L.E. Licht. 1989. Temperature and sperm incorporation in polyploid salamanders. *Science* 246:1032–1034.

Bogart, J.P., and M.W. Klemens. 1997. Hybrids and genetic interactions of mole salamanders (*Ambystoma jeffersonianum* and *A. laterale*) (Amphibia: Caudata) in New York and New England. *Am. Mus. Novit.* 3218:1–78.

Bogart, J.P., and L.E. Licht. 1986. Reproduction and the origin of polyploids in hybrid salamanders of the genus *Ambystoma*. *Can. J. Genet. Cytol.* 28:605–617.

Bogart, J.P., L.E. Licht, M.J. Oldham, and S.J. Darbyshire. 1985. Electrophoretic identification of *Ambystoma laterale* and *Ambystoma texanum* as well as their diploid and triploid interspecific hybrids (Amphibia: Caudata) on Pelee Island, Ontario. *Can. J. Zool.* 63:340–347.

Bogart, J.P., L.A. Lowcock, C.W. Zeyl, and B.K. Mable. 1987. Genome constitution and reproductive biology of the *Ambystoma* hybrid salamanders on Kelleys Island in Lake Erie. *Can. J. Zool.* 65:2188–2201.

Bohlen, J., P. Ráb, V. Slechtová, M. Rábová, D. Ritterbusch, and J. Freyhof. 2002. Hybridogeneous biotypes in spined loaches (genus *Cobitis*) in Germany with implications for the conservation of such fish complexes. Pp. 311–321 in: *Freshwater Fish Conservation,* M.J. Collares-Pereira, I. Cowx, and M.M. Coelho (eds.). Blackwell, Oxford, UK.

Bolger, D.T., and T.J. Case. 1994. Divergent ecology of sympatric clones of the asexual gecko, *Lepidodactylus lugubris. Oecologia* 100:397–405.

Bongers, A.B.J., M. Sukkel, G. Gort, J. Komen, and C.J.J. Richter. 1998. Development and use of genetically uniform strains of common carp in experimental animal research. *Lab. Anim.* 32:349–363.

Bosch, I., R.B. Rivkin, and S.P. Alexander. 1989. Asexual reproduction by oceanic planktotrophic echinoderm larvae. *Nature* 337:169–170.

Bouchard, T.J., D.T. Lykken, M. McGue, N.L. Segal, and A. Tellegen. 1990. Sources of human psychological differences: the Minnesota study of twins reared apart. *Science* 250:223–228.

Breder, C.M., and D.E. Rosen. 1966. *Modes of Reproduction in Fishes.* Natural History Press, Garden City, NY.

Briedis, A., and R.P. Elinson. 1982. Suppression of male pronuclear movement in frog eggs by hydrostatic pressure and deuterium oxide yields androgenetic haploids. *J. Exp. Zool.* 222:45–57.

Briggs, R., and T.J. King. 1952. Transplantation of living nuclei from blastula cells into enucleated frogs' eggs. *Proc. Natl. Acad. Sci. USA* 38:455–463.

Bromhall, J.D. 1975. Nuclear transplantation in the rabbit egg. *Nature* 258:719–722.

Brown, A.H.D. 1989. Genetic characterization of plant mating systems. Pp. 145–162 in: *Plant Population Genetics, Breeding and Genetic Resources,* A.H.D. Brown, M.T. Clegg, A.L. Kahler, and B.S. Weir (eds.). Sinauer, Sunderland, MA.

Brown, K.H., and G.H. Thorgaard. 2002. Mitochondrial and nuclear inheritance in an androgenetic line of rainbow trout, *Oncorhynchus mykiss. Aquaculture* 204:323–335.

Brown, W.M., M. George, and A.C. Wilson. 1979. Rapid evolution of animal mitochondrial DNA. *Proc. Natl. Acad. Sci. USA* 76:1967–1971.

Brown, W.M., and J.W. Wright. 1975. Mitochondrial DNA and the origin of parthenogenesis in whiptail lizards (genus *Cnemidophorus*). *Herpetol. Rev.* 6:70–71.

Brown, W.M., and J.W. Wright. 1979. Mitochondrial DNA analysis and the origin and relative age of parthenogenetic lizards (genus *Cnemidophorus*). *Science* 203:1247–1249.

Buchanan, G.D. 1957. Variation in litter size of nine-banded armadillos. *J. Mammal.* 38:529.

Bulger, A.J., and R.J. Schultz. 1979. Heterosis and interclonal variation in thermal tolerance in unisexual fishes. *Evolution* 33:848–859.

Bulger, A.J., and R.J. Schultz. 1982. Origins of thermal adaptation in northern vs. southern populations of a unisexual hybrid fish. *Evolution* 36:1041–1050.

Bullini, L. 1994. Origin and evolution of animal hybrid species. *Trends Ecol. Evol.* 9:422–426.

Bulmer, M.G. 1970. *The Biology of Twinning in Man*. Clarendon Press, Oxford, UK.

Buth, D.G., M.S. Gordon, I. Plaut, S.L. Drill, and L.G. Adams. 1995. Genetic heterogeneity in isogenic homozygous clonal zebrafish. *Proc. Natl. Acad. Sci. USA* 92:12367–12369.

Butlin, R. 2000. Virgin rotifers. *Trends Ecol. Evol.* 15:389–390.

Butlin, R., I. Schön, and K. Martens. 1999. Origin, age and diversity of clones. *J. Evol. Biol.* 12:1020–1022.

Byrne, J.A., D.A. Pedersen, L.L. Clepper, M. Nelson, W.G. Sanger, S. Gokhale, D.P. Wolf, and S.M. Mitalipov. 2007. Producing primate embryonic stem cells by somatic cell nuclear transfer. *Nature* 450:497–502.

Carmona, J.A., O.I. Sanjur, I. Doadrio, A. Machurdom, and R.C. Vrijenhoek. 1997. Hybridogenetic reproduction and maternal ancestry of polyploid Iberian fish: the *Tropidophoxinellus alburnoides* complex. *Genetics* 146:983–993.

Carr, D.E., and M.R. Dudash. 2003. Recent approaches into the genetic basis of inbreeding depression in plants. *Philos. Trans. R. Soc. Lond. B Biol. Sci.* 358:1071–1084.

Carter, T.C., L.C. Dunn, D.S. Falconer, H. Grüneberg, W.E. Heston, and G.D. Snell. 1952. Standardized nomenclature for inbred strains of mice. *Cancer Res.* 12:602–613.

Case, T.J. 1990. Patterns of coexistence in sexual and asexual species of *Cnemidophorus* lizards. *Oecologia* 83:220–227.

Case, T.J., and M.L. Taper. 1986. On the coexistence and coevolution of asexual and sexual competitors. *Evolution* 40:366–387.

Chan, A.W.S., T. Dominko, C.M. Luetjens, E. Neuber, C. Martinovich, L. Hewitson, C.R. Simerly, and G.P. Schatten. 2000. Clonal propagation of primate offspring by embryo splitting. *Science* 287:317–319.

Chang, M.C. 1954. Development of parthenogenetic rabbit blastocysts induced by low temperature storage of unfertilized ova. *J. Exp. Zool.* 125:127–149.

Chapman, D.D., M.S. Shivji, E. Louis, J. Sommer, H. Fletcher, and P.A. Prodöhl. 2007. Virgin birth in a hammerhead shark. *Biol. Lett.* 3:425–427.

Charlesworth, B. 1984. Androdioecy and the evolution of dioecy. *Biol. J. Linn. Soc.* 22:333–348.

Charlesworth, B. 2007. Why bother? The evolutionary genetics of sex. *Daedalus* 136(2):37–46.

Charlesworth, D., and B. Charlesworth. 1978. A model for the evolution of dioecy and gynodioecy. *Am. Nat.* 112:975–997.

Charlesworth, D., and B. Charlesworth. 1987. Inbreeding depression and its evolutionary consequences. *Annu. Rev. Ecol. Syst.* 18:237–268.

Cheng, K.C., and J.L. Moore. 1997. Genetic dissection of vertebrate processes in the zebrafish: a comparison of uniparental and two-generation screens. *Biochem. Cell Biol.* 75:525–533.

Cherfas, N.B. 1966. Natural triploidy in females of the unisexual form of silver carp (*Carassius auratus gibelio* Bloch). *Genetica* 5:16–24.

Cherfas, N.B. 1981. Gynogenesis in fishes. Pp. 255–273 in: *Genetic Bases of Fish Selection,* V.S. Kirpichnikov (ed.). Springer-Verlag, Berlin.

Cherfas, N.B., B.I. Gromelsky, O.V. Emelyanova, and A.V. Recourbratsky. 1994. Induced diploid gynogenesis and polyploidy in crucian carp, *Carassius auratus gibelio* (Bloch) X common carp, *Cyprinus carpio* L., hybrids. *Aquac. Fish. Manag.* 25:943–954.

Chesne, P., P.G. Adenot, C. Viglietta, M. Baratte, L. Boulanger, and J.P. Renard. 2002. Cloned rabbits produced by nuclear transfer from adult somatic cells. *Nat. Biotechnol.* 20:366–369.

Cibelli, J. 2007. A decade of cloning mystique. *Science* 316:990–992.

Cibelli, J.B., S.L. Stice, P.J. Golueke, J.J. Kane, J. Jerry, C. Blackwell, F.A. Ponce de León, and J.M. Robl. 1998. Cloned transgenic calves produced from nonquiescent fetal fibroblasts. *Science* 280:1256–1258.

Cimino, M.C. 1972a. Meiosis in a triploid all-female fish (*Poeciliopsis,* Poeciliidae). *Science* 175:1484–1486.

Cimino, M.C. 1972b. Egg production, polyploidization and evolution in a diploid all-female fish of the genus *Poeciliopsis. Evolution* 26:294–306.

Clanton, W. 1934. An unusual situation in the salamander *Ambystoma jeffersonianum* (Green). *Occas. Pap. Mus. Zool. Univ. Mich.* 290:1–14.

Clegg, M.T. 1980. Measuring plant mating systems. *BioScience* 30:814–818.

Clegg, M.T., and R.W. Allard. 1972. Patterns of genetic differentiation in the slender wild oat species *Avena barbata. Proc. Natl. Acad. Sci. USA* 69:1820–1824.

Cole, C.J. 1975. Evolution of parthenogenetic species of reptiles. Pp. 340–355 in: *Intersexuality in the Animal Kingdom,* R. Reinboth (ed.). Springer-Verlag, Berlin.

Cole, C.J. 1985. Taxonomy of parthenogenetic species of hybrid origin. *Syst. Zool.* 34:369–363.

Cole, C.J., and H.D. Dessauer. 1993. Unisexual and bisexual whiptail lizards of the *Cnemidophorus lemniscatus* complex (Squamata: Teiidae) of the Guiana region, South America, with descriptions of new species. *Am. Mus. Novit.* 3081:1–30.

Cole, C.J., H.C. Dessauer, and G.F. Barrowclough. 1988. Hybrid origin of a unisexual species of whiptail lizard, *Cnemidophorus neomexicanus,* in western North America: new evidence and a review. *Am. Mus. Novit.* 2905:1–38.

Cole, C.J., H.C. Dessauer, and A.L. Markezich. 1993. Missing link found: the second ancestor of *Gymnophthalmus underwoodi* (Squamata: Teiidae), a South American unisexual lizard of hybrid origin. *Am. Mus. Novit.* 3055:1–13.

Cole, C.J., H.C. Dessauer, and C.R. Townsend. 1983. Isozymes reveal hybrid origin of neotropical unisexual lizards. *Isozyme Bull.* 16:74.

Cole, C.J., H.C. Dessauer, C.R. Townsend, and M.G. Arnold. 1990. Unisexual lizards of the genus *Gymnophthalmus* (Reptilia: Teiidae) in the Neotropics: genetics, origin, and systematics. *Am. Mus. Novit.* 2994:1–29.

Cole, C.J., H.C. Dessauer, C.R. Townsend, and M.G. Arnold. 1995. *Kentropyx borckiana* (Squamata: Teiidae): a unisexual lizard of hybrid origin in the Guiana region, South America. *Am. Mus. Novit.* 3145:1–23.

Cole, K.S., and D.L.G. Noakes. 1997. Gonadal development and sexual allocation in mangrove killifish, *Rivulus marmoratus* (Pisces: Atherinomorpha). *Copeia* 1997:596–600.

Colegrave, N. 2002. Sex releases the speed limit on evolution. *Nature* 420:664–666.

Collares-Pereira, M.J. 1989. Hybridization between European cyprinids: evolutionary potential of unisexual populations. Pp. 281–288 in: *Evolution and Ecology of Unisexual Vertebrates,* R.M. Dawley and J.P. Bogart (eds.). New York State Museum, Albany.

Cordes, J.E., J.M. Walker, and R.M. Abuhteba. 1990. Genetic homogeneity in geographically remote populations of parthenogenetic *Cnemidophorus neomexicanus* (Sauria: Teiidae). *Tex. J. Sci.* 42:303–305.

Craig, S.F., L.B. Slobodkin, and G.A. Wray. 1995. The "paradox" of polyembryony. *Trends Ecol. Evol.* 10:371–372.

Craig, S.F., L.B. Slobodkin, G.A. Wray, and C.H. Biermann. 1997. The "paradox" of polyembryony: a review of the cases and a hypothesis for its evolution. *Evol. Ecol.* 11:127–143.

Crews, D., and K.T. Fitzgerald. 1980. Sexual behavior in parthenogenetic lizards *Cnemidophorus. Proc. Natl. Acad. Sci. USA* 77:499–502.

Crews, D., M. Grassman, and J. Lindzey. 1986. Behavioral facilitation of reproduction in sexual and unisexual whiptail lizards. *Proc. Natl. Acad. Sci. USA* 83:9547–9550.

Cruz, Y.P. 1981. A sterile defender morph in a polyembryonic hymenopterous parasite. *Nature* 294:446–447.

Cuellar, O. 1971. Reproduction and the mechanism of meiotic restitution in the parthenogenetic lizard *Cnemidophorus uniparens. J. Morphol.* 133:139–166.

Cuellar, O. 1974. On the origin of parthenogenesis in vertebrates: The cytogenetic factors. *Am. Nat.* 108:625–648.

Cuellar, O. 1977. Animal parthenogenesis. *Science* 197:837–843.

Cuellar, O. 1984. Histocompatibility in Hawaiian and Polynesian populations of the parthenogenetic gecko *Lepidodactylus lugubris. Evolution* 38:176–185.

Cullum, A. 1997. Comparisons of physiological performance in sexual and asexual whiptail lizards (*Cnemidophorus*): implications for the role of heterozygosity. *Am. Nat.* 150:24–47.

Dallapiccola, B., C. Stomeo, G. Ferranti, A. Di Lecce, and M. Purpura. 1985. Discordant sex in one of three monozygotic triplets. *J. Med. Genet.* 22:6–11.

Dannewitz, J., and H. Jansson. 1996. Triploid progeny from a female Atlantic salmon X brown trout hybrid backcrossed to a male brown trout. *J. Fish Biol.* 48:144–148.

Darevsky, I.S. 1966. Natural parthenogenesis in certain subspecies of Caucasian rock lizards related to *Lacerta saxicola* Eversmann. *J. Ohio Herpetol. Soc.* 5:115–152.

Darevsky, I.S. 1992. Evolution and ecology of parthenogenesis in reptiles. *Soc. Study Amphib. Reptiles Contr. Herpetol.* 9:21–39.

Darevsky, I.S., L.A. Kupriyanova, and V.V. Roshchin. 1984. A new all-female triploid species of gecko and karyological data on the bisexual *Hemidactylus frenatus* from Vietnam. *J. Herpetol.* 18:277–284.

Darevsky, I.S., L.A. Kupriyanova, and T. Uzzell. 1985. Parthenogenesis in reptiles. pp. 411–526 in: *Biology of the Reptilia,* vol. 15, C. Gans and F. Billett (eds.). Wiley, New York.

Darnell, R.M., E. Lamb, and P. Abramoff. 1967. Matroclinous inheritance and clonal structure of a Mexican population of the gynogenetic fish, *Poecilia formosa. Evolution* 21:168–173.

Darvasi, A. 1998. Experimental strategies for the genetic dissection of complex traits in animal models. *Nat. Genet.* 18:19–24.

Darwin, C.D. 1859. *On the Origin of Species by Means of Natural Selection.* Murray, London.

Darwin, C.D. 1877. *The Different Forms of Flowers on Plants of the Same Species.* Appleton, New York.

Davis, W.P., D.S. Taylor, and B.J. Turner. 1990. Field observations of the ecology and habits of mangrove rivulus (*Rivulus marmoratus*) in Belize and Florida (Teleostei: Cyprinodontiformes: Rivulidae). *Icthyol. Explor. Freshw.* 1:123–134.

Davis, W.P., D.S. Taylor, and B.J. Turner. 1995. Does the autecology of the mangrove *Rivulus* fish (*Rivulus marmoratus*) reflect a paradigm for mangrove ecosystem sensitivity? *Bull. Mar. Sci.* 57:208–214.

Dawley, R.M. 1989. An introduction to unisexual vertebrates. Pp. 1–18 in: *Evolution and Ecology of Unisexual Vertebrates,* R.M. Dawley and J.P. Bogart (eds.). New York State Museum, Albany.

Dawley, R.M. 1992. Clonal hybrids of the common laboratory fish *Fundulus heteroclitus. Proc. Natl. Acad. Sci. USA* 89:2485–2488.

Dawley, R.M., and J.P. Bogart, eds. 1989. *Evolution and Ecology of Unisexual Vertebrates.* New York State Museum, Albany.

Dawley, R.M., and K.A. Goddard. 1988. Diploid-triploid mosaics among unisexual hybrids of the minnows *Phoxinus eos* and *Phoxinus neogaeus. Evolution* 42:649–659.

Dawley, R.M., R.J. Schultz, and K.A. Goddard. 1987. Clonal reproduction and polyploidy in unisexual hybrids of *Phoxinus eos* and *Phoxinus neogaeus* (Pisces; Cyprinidae). *Copeia* 1987:275–283.

Dawley, R.M., A.M. Yeakel, K.A. Beaulieu, and K.L. Phiel. 2000. Histocompatibility analysis of clonal diversity in unisexual hybrids of the killifishes *Fundulus heteroclitus* and *Fundulus diaphanus. Can. J. Zool.* 78:923–930.

Densmore, L.D., III, C.C. Moritz, J.W. Wright, and W.M. Brown. 1989a. Mitochondrial-DNA analyses and the origin and relative age of parthenogenetic lizards (genus *Cnemidophorus*). IV. Nine *sexlineatus*-group unisexuals. *Evolution* 43:969–983.

Densmore, L.D., III, J.W. Wright, and W.M. Brown. 1989b. Mitochondrial-DNA analyses and the origin and relative age of parthenogenetic lizards (genus *Cnemidophorus*). II. *C. neomexicanus* and the *C. tesselatus* complex. *Evolution* 43:943–957.

Dessauer, H.C., and C.J. Cole. 1986. Clonal inheritance in parthenogenetic whiptail lizards: biochemical evidence. *J. Hered.* 77:8–12.

Dessauer, H.C., and C.J. Cole. 1989. Diversity between and within nominal forms of unisexual teiid lizards. Pp. 49–71 in: *Evolution and Ecology of Unisexual Vertebrates,* R.M. Dawley and J.P. Bogart (eds.). New York State Museum, Albany.

Dessauer, H.C., C.J. Cole, and C.R. Townsend. 2000. Hybridization among western whiptail lizards (*Cnemidophorus tigris*) in southwestern New Mexico: population genetics, morphology, and ecology in three contact zones. *Bull. Am. Mus. Nat. Hist.* 246:1–148.

Dessauer, H.C., T.W. Reeder, C.J. Cole, and A. Knight. 1996. Rapid screening of DNA diversity using dot-blot technology and allele-specific oligonucleotides: maternity of hybrids and unisexual clones of hybrid origin (lizards, *Cnemidophorus*). *Mol. Phylogenet. Evol.* 6:366–372.

Domes, K., R.A. Norton, M. Maraun, and S. Scheu. 2007. Re-evolution of sexuality breaks Dollo's law. *Proc. Natl. Acad. Sci. USA* 104:7139–7144.

Donnellan, S.C., and C. Moritz. 1995. Genetic diversity of bisexual and parthenogenetic populations of the tropical gecko *Nactus pelagicus* (Lacertilia, Gekkonidae). *Herpetologica* 51:140–154.

Downs, F.L. 1978. Unisexual *Ambystoma* from the Bass Islands of Lake Erie. *Occas. Pap. Mus. Zool. Univ. Mich.* 685:1–36.

Dries, L.A. 2003. Peering through the looking glass at a sexual parasite: are Amazon mollies Red Queens? *Evolution* 57:1387–1396.

Dubach, J., A. Sajewicz, and R. Pawley. 1997. Parthenogenesis in the Arafuran filesnake (*Acrochordus arafurae*). *Herpetol. Nat. Hist.* 5:11–18.

Dubois, A., and R. Günther. 1982. Klepton and synklepton: two new evolutionary systematics categories in zoology. *Zool. Jahrb. Syst.* 109:290–305.

Echelle, A.A., T.E. Dowling, C.C. Moritz, and W.M. Brown. 1989a. Mitochondrial-DNA diversity and the origin of the *Menidia clarkhubbsi* complex of unisexual fishes (Atherinidae). *Evolution* 43:984–993.

Echelle, A.A., and A.F. Echelle. 1997. Patterns of abundance and distribution among members of a unisexual-bisexual complex of fishes (Atherinidae: *Menidia*). *Copeia* 1997:249–259.

Echelle, A.A., A.F. Echelle, and C.D. Crozier. 1983. Evolution of an all-female fish, *Menidia clarkhubbsi* (Atherinidae). *Evolution* 37:772–784.

Echelle, A.A., A.F. Echelle, L.E. DeBault, and D.W. Durham. 1988. Ploidy levels in silverside fishes (Atherinidae, *Menidia*) on the Texas coast: flow-cytometric analysis of the occurrence of allotriploidy. *J. Fish Biol.* 32:835–844.

Echelle, A.A., A.F. Echelle, and D.P. Middaugh. 1989b. Evolutionary biology of the *Menidia clarkhubbsi* complex of unisexual fishes (Atherinidae): origins, clonal diversity, and mode of reproduction. Pp. 144–152 in: *Evolution and Ecology of Unisexual Vertebrates,* R.M. Dawley and J.P. Bogart (eds.). New York State Museum, Albany.

Echelle, A.A., and D.T. Mosier. 1982. *Menidia clarkhubbsi,* n. sp. (Pisces: Atherinidae), an all-female species. *Copeia* 1982:533–540.

Elinson, R.P., J.P. Bogart, L.E. Licht, and L.A. Lowcock. 1992. Gynogenetic mechanisms in polyploidy hybrid salamanders. *J. Exp. Zool.* 264:93–99.

Ellinger, M.S., D.R. King, and R.G. McKinnell. 1975. Androgenetic haploid development produced by ruby laser irradiation of anuran ova. *Radiat. Res.* 62:117–122.

Engeler, B., and H.-U. Reyer. 2001. Choosy females and indiscriminate males: mate choice in mixed populations of sexual and hybridogenetic water frogs (*Rana lessonae, Rana esculenta*). *Behav. Ecol.* 12:600–606.

Ensminger, M.E. 1980. *Dairy Cattle Science.* 2nd ed. Interstate, Danville, IL.

Ercanbrack, S.K., and A.D. Knight. 1993. Ten-year linear trends in reproduction and wool production among inbred and noninbred lines of Rambouillet, Targhee, and Columbia sheep. *J. Anim. Sci.* 71:341–354.

Fangerau, H. 2005. Can artificial parthenogenesis sidestep ethical pitfalls in human therapeutic cloning? An historical perspective. *J. Med. Ethics* 31:733–735.

Felsenstein, J. 1974. The evolutionary advantage of recombination. *Genetics* 78:737–756.

Fernandez, M. 1909. Bieträge zur embryologie der Gurtelziere, 1. *Morph. Jahrb. Bd.* 39:302–333.

Fischer, E.A., and C.W. Petersen. 1987. The evolution of sexual patterns in the seabasses. *BioScience* 37:482–489.

Fisher, R.A. 1930. *The Genetical Theory of Natural Selection.* Clarendon Press, Oxford, UK.

Fraga, M.F., E. Ballestar, M.F. Paz, S. Ropero, F. Setien, M.L. Ballestar, D. Heine-Suñer, J.C. Cigudosa, M. Urioste, J. Benitez, et al. 2005. Epigenetic differences arise during the lifetime of monozygotic twins. *Proc. Natl. Acad. Sci. USA* 102:10604–10609.

Frankham, R., J.D. Ballou, and D.A. Briscoe. 2002. *Introduction to Conservation Genetics.* Cambridge University Press, Cambridge, UK.

Frick, N.T., and P.A. Wright. 2002. Nitrogen metabolism and excretion in the mangrove killifish *Rivulus marmoratus.* II. Significant ammonia volatilization in a teleost during air exposure. *J. Exp. Biol.* 205:91–100.

Fritts, T.H. 1969. The systematics of the parthenogenetic lizards of the *Cnemidophorus cozumela* complex. *Copeia* 1969:519–535.

Frost, D.R., and J.W. Wright. 1988. The taxonomy of uniparental species, with special reference to parthenogenetic *Cnemidophorus* (Squamata: Teiidae). *Syst. Zool.* 37:200–209.

Fu, J., R.D. MacCulloch, R.W. Murphy, I.S. Darevsky, L.A. Kupriyanova, and F. Danielyan. 1998. The parthenogenetic rock lizard *Lacerta unisexualis:* an example of limited polymorphism. *J. Mol. Evol.* 46:127–130.

Fu, J., R.W. Murphy, and I.S. Darevsky. 2000. Divergence of the cytochrome *b* gene in the *Lacerta raddei* complex and its parthenogenetic daughter species: evidence for recent multiple origins. *Copeia* 2000:432–440.

Gabriel, W., M. Lynch, and R. Bürger. 1993. Muller's ratchet and mutational meltdown. *Evolution* 47:1744–1757.

Galbreath, G.J. 1985. The evolution of monozygotic polyembryony in *Dasypus*. Pp. 243–246 in: *The Evolution and Ecology of Armadillos, Sloths, and Vermilinguas,* G.G. Montgomery (ed.). Smithsonian Institution Press, Washington, DC.

Galbreath, P.F., K.J. Adams, P.A. Wheeler, and G.H. Thorgaard. 1997. Clonal Atlantic salmon X brown trout hybrids produced by gynogenesis. *J. Fish Biol.* 50:1025–1033.

Galbreath, P.F., and G.H. Thorgaard. 1995. Sexual maturation and fertility of diploid and triploid Atlantic salmon X brown trout triploid hybrids. *Can. J. Fish. Aquat. Sci.* 51(Suppl. 1):16–24.

Galli, C., I. Lagutina, G. Crotti, S. Colleoni, P. Turini, N. Ponderato, R. Duchi, and G. Lazzari. 2003a. Pregnancy: a cloned horse foal born to its dam twin. *Nature* 424:635.

Galli, C, I. Lagutina, and G. Lazzari. 2003b. Introduction to cloning by nuclear transplantation. *Cloning Stem Cells* 5:223–232.

Gatenby, J.B. 1918. Polyembryony in parasitic Hymenoptera—a review. *Q. J. Microsc. Sci.* 63:175–214.

Georgiades, P., M. Watkins, G.J. Burton, and A.C. Ferguson-Smith. 2001. Roles for genomic imprinting and the zygotic genome in placental development. *Proc. Natl. Acad. Sci. USA* 98:4522–4527.

Ghiselin, M.T. 1969. The evolution of hermaphroditism among animals. *Q. Rev. Biol.* 44:189–208.

Ghiselin, M.T. 1974. *The Economy of Nature and the Evolution of Sex.* University of California Press, Berkeley.

Gillespie, J.H. 2000. Genetic drift in an infinite population: the pseudohitchhiking model. *Genetics* 155:909–919.

Gillespie, L.L., and J.B. Armstrong. 1979. Induction of triploid and gynogenetic diploid axolotls (*Ambystoma mexicanum*) by hydrostatic pressure. *J. Exp. Zool.* 210:117–121.

Gillespie, L.L., and J.B. Armstrong. 1980. Production of androgenetic diploid axolotls by suppression of first cleavage. *J. Exp. Zool.* 213:423–425.

Gillespie, L.L., and J.B. Armstrong. 1981. Suppression of first cleavage in the Mexican axolotl (*Ambystoma mexicanum*) by heat shock or hydrostatic pressure. *J. Exp. Zool.* 218:441–445.

Giron, D., D.W. Dunn, I.C.W. Hardy, and M.R. Strand. 2004. Aggression by polyembryonic wasp soldiers correlates with kinship but not resource competition. *Nature* 430:676–679.

Gleeson, S.K., A.B. Clark, and L.A. Dugatkin. 1994. Monozygotic twinning: an evolutionary hypothesis. *Proc. Natl. Acad. Sci. USA* 91:11363–11367.

Goddard, K.A., and R.M. Dawley. 1990. Clonal inheritance of a diploid nuclear genome by a hybrid freshwater minnow (*Phoxinus eos-neogaeus,* Pisces: Cyprinidae). *Evolution* 44:1052–1065.

Goddard, K.A., R.M. Dawley, and T.E. Dowling. 1989. Origin and genetic relationships of diploid, triploid, and diploid-triploid mosaic biotypes in the *Phoxinus eos-neogaeus* unisexual complex. Pp. 268–280 in: *Evolution and Ecology of Unisexual Vertebrates,* R.M. Dawley and J.P. Bogart (eds.). New York State Museum, Albany.

Goddard, K.A., and R.J. Schultz. 1993. Aclonal reproduction by polyploid members of the clonal hybrid species *Phoxinus eos-neogaeus* (Cyprinidae). *Copeia* 1993:650–660.

Godfray, H.C.J. 1994. *Parasitoids: Behavioral and Evolutionary Ecology.* Princeton University Press, Princeton, NJ.

Gómez, M.C., C.E. Pope, A. Giraldo, L.A. Lyons, R.F. Harris, A.L. King, A. Cole, R.A. Godke, and B.L. Dresser. 2004. Birth of African wildcat cloned kittens born from domestic cats. *Cloning Stem Cells* 6:247–258.

Good, D.A., and J.W. Wright. 1984. Allozymes and the hybrid origin of the parthenogenetic lizard *Cnemidophorus exsanguis. Experientia* 40:1012–1014.

Goodwillie, C., S. Kalisz, and C.G. Eckert. 2005. The evolutionary enigma of mixed mating systems in plants: occurrence, theoretical explanations, and empirical evidence. *Annu. Rev. Ecol. Evol. Syst.* 36:47–79.

Gorokhova, E., T.E. Dowling, L.J. Weider, T.J. Crease, and J.J. Elser. 2002. Functional and ecological significance of rDNA intergenic spacer variation in a clonal organism under divergent selection for production rate. *Proc. R. Soc. Lond. B Biol. Sci.* 269:2373–2379.

Graf, J.-D., F. Karch, and M.-C. Moreillon. 1977. Biochemical variation in the *Rana esculenta* complex: a new hybrid form related to *Rana perezi* and *Rana ridibunda. Experientia* 33:1582–1584.

Graf, J.-D., and M. Polls Pelaz. 1989. Evolutionary genetics of the *Rana esculenta* complex. Pp. 289–302 in: *Evolution and Ecology of Unisexual Vertebrates,* R.M. Dawley and J.P. Bogart (eds.). New York State Museum, Albany.

Grant, V. 1958. The regulation of recombination in plants. *Cold Spring Harbor Symp. Quant. Biol.* 23:337–363.

Grant, V. 1981. *Plant Speciation.* 2nd ed. Columbia University Press, New York.

Grbic, M., L.M. Nagy, and M.R. Strand. 1998. Development of polyembryonic insects: a major departure from typical insect embryogenesis. *Dev. Genes Evol.* 208:69–81.

Green, R.F., and D.L.G. Noakes. 1995. Is a little bit of sex as good as a lot? *J. Theor. Biol.* 174:87–96.

Grizzle, J.M., and A. Thiyagarajah. 1987. Skin histology of *Rivulus ocellatus marmoratus,* apparent adaptation for aerial respiration. *Copeia* 1987:237–240.

Groot, T.V.M., E. Bruins, and J.A.J. Breeuwer. 2003. Molecular genetic evidence for parthenogenesis in the Burmese python, *Python molurus bivittatus. Heredity* 90:130–135.

Gurdon, J.B., and J.A. Byrne. 2003. The first half-century of nuclear transplantation. *Proc. Natl. Acad. Sci. USA* 100:8048–8052.

Gurdon, J.B., R.A. Laskey, and O.R. Reeves. 1975. The developmental capacity of nuclei transplanted from keratinized skin cells of adult frogs. *J. Embryol. Exp. Morphol.* 34:93–112.

Gyllensten, U.B., D. Wharton, A. Josefsson, and A.C. Wilson. 1991. Paternal inheritance of mitochondrial DNA in mice. *Nature* 352:255–257.

Haag-Liautard, C., M. Dorris, X. Maside, S. Macaskill, D.L. Halligan, B. Charlesworth, and P.D. Keightley. 2007. Direct estimation of per nucleotide and genomic deleterious mutation rates in *Drosophila. Nature* 445:82–85.

Haccou, P., and M.V. Schneider. 2004. Modes of reproduction and the accumulation of deleterious mutations with multiplicative fitness effects. *Genetics* 166:1093–1104.

Halkett, F., J.-C. Simon, and F. Balloux. 2005. Tackling the population genetics of clonal and partially clonal organisms. *Trends Ecol. Evol.* 20:194–201.

Hall, W.P. 1970. Three probable cases of parthenogenesis in reptiles (Agamidae, Chamaeleonidae, Gekkonidae). *Experientia* 26:1271–1273.

Hamilton, W.D. 1964. The genetical evolution of social behavior. *J. Theor. Biol.* 7:1–52.

Hamilton, W.D. 1980. Sex versus non-sex versus parasite. *Oikos* 35:282–290.

Hamlett, G.W.D. 1933. Polyembryony in the armadillo: genetic or physiological? *Q. Rev. Biol.* 8:348–358.

Hamrick, J.L., and R.W. Allard. 1972. Microgeographical variation in allozyme frequencies in *Avena barbata. Proc. Natl. Acad. Sci. USA* 69:2100–2104.

Hanley, K.A., D.T. Bolger, and T.J. Case. 1994. Comparative ecology of sexual and asexual gecko species (*Lepidodactylus*) in French Polynesia. *Evol. Ecol.* 8:438–454.

Hanley, K.A., R.N. Fisher, and T.J. Case. 1995. Lower mite infestations in an asexual gecko compared with its sexual ancestors. *Evolution* 49:418–426.

Hao, Y.-H., L.X. Lai, Z.H. Liu, G.S. Im, D. Wax, M. Samuel, C.N. Murphy, P. Sutovsky, and R.S. Prather. 2006. Developmental competence of porcine parthenogenetic embryos relative to embryonic chromosomal abnormalities. *Mol. Reprod. Dev.* 73:77–82.

Hardy, I.C.W. 1995a. Protagonists of polyembryony. *Trends Ecol. Evol.* 10:179–180.

Hardy, I.C.W. 1995b. Protagonists of polyembryony. Reply to Craig et al. from Hardy. *Trends Ecol. Evol.* 10:372.

Hardy, L.M., C.J. Cole, and C.R. Townsend. 1989. Parthenogenetic reproduction in the neotropical unisexual lizard, *Gymnophthalmus underwoodi* (Reptilia: Teiidae). *J. Morphol.* 201:215–234.

Harmer, S.F. 1890. On the origin of the embryos in the ovicells of cyclostomatous Polyzoa. *Proc. Cambridge Phil. Soc.* 7:48–55.

Harmer, S.F. 1893. On the occurrence of embryonic fission in cyclostomatous Polyzoa. *Q. J. Microsc. Sci.* 34:199–241.

Harrington, R.W., Jr. 1961. Oviparous hermaphroditic fish with internal self-fertilization. *Science* 134:1749–1750.

Harrington, R.W., Jr. 1963. Twenty-four hour rhythms of internal self-fertilization and of oviposition by hermaphrodites of *Rivulus marmoratus*. *Physiol. Zool.* 36:325–331.

Harrington, R.W., Jr. 1967. Environmentally controlled induction of primary male gonochorists from eggs of the self-fertilizing hermaphroditic fish, *Rivulus marmoratus* Poey. *Biol. Bull.* 132:174–199.

Harrington, R.W., Jr. 1968. Delimitation of the thermolabile phenocritical period of sex determination from eggs of the self-fertilizing hermaphroditic fish *Rivulus marmoratus*. *Physiol. Zool.* 41:447–460.

Harrington, R.W., Jr. 1971. How ecological and genetic factors interact to determine when self-fertilizing hermaphrodites of *Rivulus marmoratus* change into functional secondary males, with a reappraisal of the modes of intersexuality among fishes. *Copeia* 1971:389–432.

Harrington, R.W., Jr., and K.D. Kallman. 1968. The homozygosity of clones of the self-fertilizing hermaphrodite fish *Rivulus marmoratus* Poey (Cyprinodontidae, Atheriniformes). *Am. Nat.* 102:337–343.

Hedges, S.B., J.P. Bogart, and L.R. Maxson. 1992. Ancestry of unisexual salamanders. *Nature* 356:708–710.

Hedrick, P.W. 2000. *Genetics of Populations*. Jones and Bartlett, Sudbury, MA.

Heethoff, M., K. Domes, M. Laumann, M. Maraun, R.A. Norton, and S. Scheu. 2007. High genetic divergences indicate ancient separation of parthenogenetic lineages of the oribatid mite *Platynothrus peltifer* (Acari, Oribatida). *J. Evol. Biol.* 20:392–402.

Helfman, G.S., B.B. Collette, and D.E. Lacy. 1997. *The Diversity of Fishes*. Blackwell, Malden, MA.

Hellriegel, B., and H.-U. Reyer. 2000. Factors influencing the composition of mixed populations of a hemiclonal hybrid and its sexual host. *J. Evol. Biol.* 13:906–918.

Hernandez-Gallegos, O., N. Manriquez-Moran, F.R. Mendez, M. Villagran, and O. Cuellar. 1998. Histocompatibility in parthenogenetic lizards of the *Cnemidophorus cozumela* complex from the Yucatan Peninsula of Mexico. *Biogeographica* 74: 117–124.

Hertwig, O. 1911. Die radium krankheit tierischer keim zellen. *Arch. Mikrosk. Anat.* 77:1–97.

Heyman, Y., X. Vignon, P. Chesne, D. Le Bourhis, J. Marchal, and J.P. Renard. 1998. Cloning in cattle: from embryo splitting to somatic nuclear transfer. *Reprod. Nutr. Dev.* 38:595–603.

Hickey, D.A., and M.R. Rose. 1988. The role of gene transfer in the evolution of eukaryotic sex. Pp. 161–175 in: *The Evolution of Sex,* R.E. Michod and B.R. Levin (eds.). Sinauer, Sunderland, MA.

Hill, W.G., and A. Robertson. 1966. The effect of linkage on limits to artificial selection. *Genet. Res.* 8:269–294.

Hotz, H., G. Mancino, S Bucci-Innocenti, M. Ragghianti, L. Berger, and T. Uzzell. 1985. *Rana ridibunda* varies geographically in inducing clonal gametogenesis in interspecies hybrids. *J. Exp. Zool.* 236:199–210.

Hotz, H., R.D. Semlitsch, E. Gutmann, G.-D. Guex, and P. Beerli. 1999. Spontaneous heterosis in larval life-history traits of hemiclonal frog hybrids. *Proc. Natl. Acad. Sci. USA* 96:2171–2176.

Hotz, H., and T. Uzzell. 1983. Interspecific hybrids of *Rana esculenta* without germ line exclusion of a parental genome. *Experientia* 39:538–540.

Howard, L.O. 1906. Polyembryony and the fixing of sex. *Science* 24:810–818.

Hubbs, C.L. and D.E.S. Brown. 1929. Materials for a distributional study of Ontario fishes. *Trans. Royal Can. Inst.* 17:1–56.

Hubbs, C.L., and L.C. Hubbs. 1932. Apparent parthenogenesis in nature, in a form of fish of hybrid origin. *Science* 76:628–630.

Hubbs, C.L., and L.C. Hubbs. 1946. Breeding experiments with the invariably female, strictly matroclinous fish, *Mollienesia formosa. Genetics* 31:218.

Huber, J.H. 1992. *Review of Rivulus: Ecobiogeography, Relationships.* Museum National d'Histoire Naturelle, Paris.

Hughes, R.N. 1989. *A Functional Biology of Clonal Animals.* Chapman and Hall, London.

Hughes, R.N., M.E. D'Amato, J.D.D. Bishop, G.R. Carvalho, S.F. Craig, L.J. Hansson, M.A. Harley, and A.J. Pemberton. 2005. Paradoxical polyembryony? Embryonic cloning in an ancient order of marine bryozoans. *Biol. Lett.* 1:178–180.

Iguchi, K., N. Matsubara, and H. Hakoyama. 2001. Behavioral individuality assessed from two strains of cloned fish. *Anim. Behav.* 61:351–356.

Illmensee, K., and P.C. Hoppe. 1981. Nuclear transplantation in *Mus musculus:* developmental potential of nuclei from preimplantation embryos. *Cell* 23:9–18.

Itono, M., K. Morishima, T. Fujimoto, E. Bando, E. Yamaha, and K. Arai. 2006. Premeiotic endomitosis produces diploid eggs in the natural clone loach, *Misgurnus anguillicaudatus* (Teleostei: Cobitidae). *J. Exp. Zool.* 305A:513–523.

Jaeckle, W.B. 1994. Multiple modes of asexual reproduction by tropical and subtropicl sea star larvae: an unusual adaptation for genet dispersal and survival. *Biol. Bull.* 186:62–71.

Janko, K., J. Bohlen, D. Lamatsch, M. Flajshans, J.T. Epplen, P. Ráb, P. Kotlík, and V. Slechtová. 2007. The gynogenetic reproduction of diploid and triploid hybrid spined loaches (*Cobitis:* Teleostei), and their ability to establish successful clonal lineages—on the evolution of polyploidy in asexual vertebrates. *Genetica* 131:185–194.

Janko, K., M.A. Culling, P. Ráb, and P. Kotlík. 2005. Ice age cloning—comparison of the Quaternary evolutionary histories of sexual and clonal forms of spiny loaches (*Cobitis:* Teleostei) using the analysis of mitochondrial DNA variation. *Mol. Ecol.* 14:2991–3004.

Janko, K., P. Kotlik, and P. Ráb. 2003. Evolutionary history of asexual hybrid loaches (*Cobitis:* Teleostei) inferred from phylogenetic analysis of mitochondrial DNA variation. *J. Evol. Biol.* 16:1280–1287.

Jarne, P., and J.R. Auld. 2006. Animals mix it up too: the distribution of self-fertilization among hermaphroditic animals. *Evolution* 60:1816–1824.

Johnson, K.R., and J.M. Wright. 1986. Female brown trout X male Atlantic salmon hybrids produce gynogens and triploids when backcrossed to male Atlantic salmon. *Aquaculture* 57:345–358.

Judson, O.P., and B.B. Normark. 1996. Ancient asexual scandals. *Trends Ecol. Evol.* 11:41–46.

Kallman, K.D. 1960. Dosage and additive effects of histocompatibility genes in the teleost *Xiphophorus maculatus. Ann. N.Y. Acad. Sci.* 87:10–43.

Kallman, K.D. 1961. Genetic homogeneity of a small isolated population of viviparous fish as revealed by tissue transplantation. *Am. Zool.* 1:455.

Kallman, K.D. 1962a. Population genetics of the gynogenetic teleost, *Mollienesia formosa* (Girard). *Evolution* 16:497–504.

Kallman, K.D. 1962b. Gynogenesis in the teleost, *Mollienesia formosa* (Girard), with a discussion of the detection of parthenogenesis in vertebrates by tissue transplantation. *J. Genet.* 58:7–21.

Kallman, K.D. 1964. Homozygosity in a gynogenetic fish—*Poecilia formosa. Genetics* 50:260–262.

Kallman, K.D. 1975. Genetics of tissue transplantation in Teleostei. *Transplant. Proc.* 2:263–271.

Kallman, K.D., and M. Gordon. 1958. Genetics of fin transplantation in xiphophorin fishes. *Ann. N.Y. Acad. Sci.* 73:599–610.

Kallman, K.D., and R.W. Harrington Jr. 1964. Evidence for the existence of homozygous clones in the self-fertilizing hermaphroditic teleost *Rivulus marmoratus* (Poey). *Biol. Bull.* 126:101–114.

Kan, N.G., I.A. Martyrosyan, I.S. Darevsky, F.D. Danielyan, M.S. Arakelyan, A.V. Aslanyan, V.V. Grechko, O.N. Tokarskaya, and A.P. Ryskov. 2000. Detection of genetically unstable loci in parthenogenetic families of lizards of the *Lacerta* genus by DNA fingerprinting. *Mol. Biol.* 34:707–711.

Katheriner, L. 1904. Ueber die Entwicklung von *Gyrodactyluys elegans* v. Nrdm. *Zool. Jahrb.* 70:519–550.

Kato, Y., T. Tani, Y. Sotomaru, K. Kurokawa, J. Kato, H. Doguchi, H. Yasue, and Y. Tsunoda. 1998. Eight calves cloned from somatic cells of a single adult. *Science* 282:2095–2098.

Kaufman, M.H. 1983. *Early Mammalian Development: Parthenogenetic Studies.* Cambridge University Press, Cambridge, UK.

Kawecki, T.J. 1988. Unisexual/bisexual breeding complexes in Poeciliidae: why do males copulate with unisexual females? *Evolution* 42:1018–1023.

Kearney, M. 2005. Hybridization, glaciation and geographical parthenogenesis. *Trends Ecol. Evol.* 20:495–502.

Kearney, M. 2006. Response to Lundmark: polyploidization, hybridization and geographical parthenogenesis. *Trends Ecol. Evol.* 21:10.

Kearney, M., M.J. Blacket, J.L. Strasburg, and C. Moritz. 2006. Waves of parthenogenesis in the desert: evidence for parallel loss of sex in a grasshopper and a gecko from Australia. *Mol. Ecol.* 15:1743–1748.

Kearney, M., and R. Shine. 2004a. Developmental success, stability, and plasticity in closely related parthenogenetic and sexual lizards (*Heteronotia,* Gekkonidae). *Evolution* 58:1560–1572.

Kearney, M., and R. Shine. 2004b. Morphological and physiological correlates of hybrid parthenogenesis. *Am. Nat.* 164:803–813.

Kearney, M., and R. Shine. 2005. Lower fecundity in parthenogenetic geckos than sexual relatives in the Australian arid zone. *J. Evol. Biol.* 18:609–618.

Kearney, M., R. Wahl, and K. Autumn. 2005. Increased capacity for sustained locomotion at low temperature in parthenogenetic geckos of hybrid origin. *Physiol. Biochem. Ecol.* 78:316–324.

Keegan-Rogers, V. 1984. Unfamiliar-female advantage among clones of unisexual fish (*Poeciliopsis,* Poeciliidae). *Copeia* 1984:169–174.

Keegan-Rogers, V., and R.J. Schultz. 1984. Differences in courtship aggression among six clones of unisexual fish. *Anim. Behav.* 32:1040–1048.

Keegan-Rogers, V., and R.J. Schultz. 1988. Sexual selection among clones of unisexual fish (*Poeciliopsis,* Poeciliidae): genetic factors and rare-female advantage. *Am. Nat.* 132:846–868.

Kim, I.-S., and E.-H. Lee. 1990. Diploid-triploid hybrid complex of the spined loach *Cobitis sinensis* and *C. longicorpus* (Pisces: Cobitidae). *Korean J. Ichthyol.* 7:71–78.

Kim, I.-S., and E.-H. Lee. 2000. Hybridization experiment of diploid-triploid cobitid fishes, *Cobitis sinensis-longicorpus* complex (Pisces, Cobitidae). *Folia Zool.* 49:17–22.

Kim, K., P. Lerou, A. Yabuuchi, C. Lengerke, K. Ng, J. West, A. Kirby, M.J. Daly, and G.Q. Daley. 2007. Histocompatible embryonic stem cells by parthenogenesis. *Science* 315:482–486.

Kim, K.-R., K. Kwon, J.Y. Joung, K.S. Kim, A.G. Ayala, and J.Y Ro. 2002. True hermaphroditism and mixed gonadal dysgenesis in young children: a clinicopathologic study of 10 cases. *Mod. Pathol.* 15:1013–1019.

Kim, M.K., G. Jang, H.J. Oh, F. Yuda, H.J. Kim, W.S. Hwang, M.S. Hossein, J.J. Kim, N.S. Shin, S.K. Kang, et al. 2007. Endangered wolves cloned from adult somatic cells. *Cloning Stem Cells* 9:130–137.

King, T.J., and R. Briggs. 1955. Changes in the nuclei of differentiating gastrula cells, as demonstrated by nuclear transplantation. *Proc. Natl. Acad. Sci. USA* 41:321–325.

Kirkendall, L.R., and N.C. Stenseth. 1990. Ecological and evolutionary stability of sperm-dependent parthenogenesis: effects of partial niche overlap between sexual and asexual females. *Evolution* 44:698–714.

Kitiyanant, Y., J. Saikhun, B. Chaisalee, K.L. White, and K. Pavasuthipaisit. 2001. Somatic cell cloning in buffalo (*Bubalus bubalis*): effects of interspecies cytoplasmic recipients and activation procedures. *Cloning Stem Cells* 3:97–104.

Kizirian, D.A., and C.J. Cole. 1999. Origin of the unisexual lizard *Gymnophthalmus underwoodi* (Gymnophthalmidae) inferred from mitochondrial DNA nucleotide sequences. *Mol. Phylogenet. Evol.* 11:394–400.

Klicka, J., and R.M. Zink. 1997. The importance of recent Ice Ages in speciation: a failed paradigm. *Science* 277:1666–1669.

Koh, C, R. Lanza, S. Soker, J. Yoo, and A. Atala. 2004. Parthenogenesis: a novel source of stem cells for reconstruction. *Urol. Reprod. Surg.* 199:S101–S102.

Koide, T., K. Moriwaki, K. Ikeda, H. Niki, and T. Shiroishi. 2000. Multi-phenotype behavioral characterization of inbred strains derived from wild stocks of *Mus musculus*. *Mamm. Genome* 11:664–670.

Kolberg, R. 1993. Human embryo cloning reported. *Science* 262:652–653.

Komen, J.A., B.J. Bongers, C.J.J. Richter, W.B. van Muiswinkel, and E.A. Huisman. 1991. Gynogenesis in common carp (*Cyprinus carpio* L.). II. The production of homozygous gynogenetic clones and F1 hybrids. *Aquaculture* 92:127–142.

Komen, J., E.H. Eding, A.B.J. Bongers, and C.J.J. Richter. 1993. Gynogenesis in common carp (*Cyrpinus carpio*). IV. Growth, phenotypic variation and gonal differentiation in normal and methyltestosterone-treated homozygous clones and F(1) hybrids. *Aquaculture* 111:271–280.

Kondo, R., Y. Satta, E.T. Matsuura, H. Ishiwa, N. Takahata, and S.I. Chigusa. 1990. Incomplete maternal transmission of mitochondrial DNA in *Drosophila*. *Genetics* 126:657–663.

Kondrashov, A.S. 1988. Deleterious mutations and the evolution of sexual reproduction. *Nature* 336:435–440.

Kondrashov, A.S. 1993. Classification of hypotheses on the advantage of amphimixis. *J. Hered.* 84:372–387.

Kono, T. 2006. Genomic imprinting is a barrier to parthenogenesis in mammals. *Cytogenet. Genome Res.* 113:31–35.

Kono, T., Y. Obata, Q. Wu, K. Niwa, Y. Ono, Y. Yamamoto, E.S. Park, J.S. Seo, and H. Ogawa. 2004. Birth of parthenogenetic mice that can develop into adulthood. *Nature* 428:860–864.

Kraus, F. 1985. Unisexual salamander lineages in northwestern Ohio and southern Michigan: a study of the consequences of hybridization. *Copeia* 1985:309–324.

Kraus, F. 1989. Constraints on the evolutionary history of the unisexual salamanders of the *Ambystoma laterale-texanum* complex as revealed by mitochondrial DNA analysis. Pp. 218–227 in: *Evolution and Ecology of Unisexual Vertebrates*, R.M. Dawley and J.P. Bogart (eds.). New York State Museum, Albany.

Kraus, F. 1995. The conservation of unisexual vertebrate populations. *Conserv. Biol.* 9:956–959.

Kraus, F., P.K. Ducey, P. Moler, and M.M. Miyamoto. 1991. Two new triparental unisexual *Ambystoma* from Ohio and Michigan. *Herpetologica* 47:429–439.

Kraus, F., and M.M. Miyamoto. 1990. Mitochondrial genotype of a unisexual salamander of hybrid origin is unrelated to either of its nuclear haplotypes. *Proc. Natl. Acad. Sci. USA* 87:2235–2238.

Kraus, F., and J.W. Petranka. 1989. A new sibling species of *Ambystoma* from the Ohio River drainage. *Copeia* 1989:94–110.

Kristensen, H. 1970. Competition in three cyprinodont fish species in the Netherlands Antilles. *Stud. Fauna Curacao Carib. Isl.* 32:82–101.

Kubota, C., H. Yamakuchi, J. Todoroki, K. Mizoshita, N. Tabara, M. Barber, and X. Yang. 2000. Six clone calves produced from adult fibroblast cells after long-term culture. *Proc. Natl. Acad. Sci. USA* 97:990–995.

Laale, H.W. 1984. Polyembryony in teleostean fishes—double monstrosities and triplets. *J. Fish Biol.* 24:711–719.

Ladle, R.J. 1992. Parasites and sex: catching the Red Queen. *Trends Ecol. Evol.* 7:405–408.

Lamatsch, D.K., I. Nanda, I. Schlupp, J.T. Epplen, M. Schmid, and M. Schartl. 2004. Distribution and stability of supernumerary microchromosomes in natural populations of the Amazon molly, *Poecilia formosa*. *Cytogenet. Genome Res.* 106:189–194.

Lamatsch, D.K., M. Schmid, and M. Schartl. 2002. A somatic mosaic of the gynogenetic Amazon molly. *J. Fish Biol.* 60:1417–1422.

Lampert, K.P., D.K. Lamatsch, J.T. Epplen, and M. Schartl. 2005. Evidence for a monophyletic origin of triploid clones of the Amazon molly, *Poecilia formosa*. *Evolution* 59:881–889.

Lampert, K.P., D.K. Lamatsch, P. Fischer, J.T. Epplen, I. Nanda, M. Schmid, and M. Schartl. 2007. Automictic reproduction in interspecific hybrids of poeciliid fish. *Curr. Biol.* 17:1948–1953.

Lanza, R.P., J.B. Cibelli, F. Diaz, C.T. Moraes, P.W. Farin, C.E. Farin, C.J. Hammer, M.D. West, and P. Damiani. 2000a. Cloning of an endangered species (*Box gaurus*) using interspecies nuclear transfer. *Cloning* 2:79–90.

Lanza, R.P., J.B. Cibelli, D. Faber, R.W. Sweeney, B. Henderson, W. Nevala, M.D. West, and P.J. Wettstein. 2001. Cloned cattle can be healthy and normal. *Science* 294:1893–1894.

Lanza, R.P., B.L. Draper, and P. Damiani. 2000b. Cloning Noah's ark. *Sci. Am.* 283:84–89.

Laughlin, T.L., B.A. Lubinski, E.-H. Park, D.S. Taylor, and B.S. Turner. 1995. Clonal stability and mutation in the self-fertilizing hermaphroditic fish, *Rivulus marmoratus. J. Hered.* 86:399–402.

Lee, B.C., M.K. Kim, G. Jang, H.J. Oh, F. Yuda, H.J. Kim, M.S. Hossein, J.J. Kim, S.K. Kang, G. Schatten, et al. 2005. Dogs cloned from adult somatic cells. *Nature* 436:641–642.

Lee, J.-S., M. Miya, Y.-S. Lee, C.G. Kim, E.-H. Park, Y. Aoki, and M. Nishida. 2001. The complete DNA sequence of the mitochondrial genome of the self-fertilizing fish *Rivulus marmoratus* (Cyprinodontiformes, Rivulidae) and the first description of duplication of a control region in fish. *Gene* 280:1–7.

Lee, K.Y., H. Huang, B. Ju, Z. Yang, and S. Lin. 2002. Cloned zebrafish by nuclear transfer from long-term-cultured cells. *Nat. Biotechnol.* 20:795–799.

Lenk, P., B. Eidenmueller, H. Staudter, R. Wicker, and M. Wink. 2005. A parthenogenetic *Varanus. Amphib.-Reptilia* 26:507–514.

Leonard, J.L. 2006. Sexual selection: lessons from hermaphroditic mating systems. *Integr. Comp. Biol.* 46:349–367.

Leslie, J.F., and R.C. Vrijenhoek. 1978. Genetic dissection of clonally inherited genomes of *Poeciliopsis*. I. Linkage analysis and preliminary assessment of deleterious gene loads. *Genetics* 90:801–811.

Leslie, J.F., and R.C. Vrijenhoek. 1980. Consideration of Muller's ratchet mechanism through studies of genetic linkage and genomic compatibilities in clonally reproducing *Poeciliopsis. Evolution* 34:1105–1115.

Levine, H., III. 1999. *Genetic Engineering: A Reference Handbook.* ABC-CLIO, Santa Barbara, CA.

Levitan, D.R., and C. Petersen. 1995. Sperm limitation in the sea. *Trends Ecol. Evol.* 10:228–231.

Li, Z., X. Sun, J. Chen, X. Liu, S.M. Wisely, Q. Zhou, J.P. Renard, G.H. Leno, and J.F. Engelhardt. 2006. Cloned ferrets produced by somatic cell nuclear transfer. *Dev. Biol.* 293:439–448.

Licht, L.W. 1989. Reproductive parameters of unisexual *Ambystoma* on Pelee Island, Ontario. Pp. 209–217 in: *Evolution and Ecology of Unisexual Vertebrates,* R.M. Dawley and J.P. Bogart (eds.). New York State Museum, Albany.

Lima, N.R.W., C.J. Kobak, and R.C. Vrijenhoek. 1996. Evolution of sexual mimicry in sperm-dependent all-female forms of *Poeciliopsis* (Pisces: Poeciliidae). *J. Evol. Biol.* 9:185–203.

Lin, H.-C., and W.A. Dunson. 1995. An explanation of the high strain diversity of a self-fertilizing hermaphroditic fish. *Ecology* 76:593–605.

Litwiller, S.L., M. O'Donnell, and P.A. Wright. 2006. Rapid increase in the partial pressure of NH_3 on the cutaneous surface of air-exposed mangrove killifish, *Rivulus marmoratus*. *J. Exp. Biol.* 209:1737–1745.

Liu, C.-T, C.-H. Chen, S.-P. Cheng, and J.-C. Ju. 2002. Parthenogenesis of rabbit oocytes activated by different stimuli. *Anim. Reprod. Sci.* 70:267–276.

Lively, C.M., C. Craddock, and R.C. Vrijenhoek. 1990. The red queen hypothesis supported by parasitism in sexual and clonal fish. *Nature* 344:864–866.

Lloyd, D.G. 1979. Some reproductive factors affecting the selection of self-fertilization in plants. *Am. Nat.* 113:67–79.

Loeb, J. 1899. On the nature of the process of fertilization and the artificial production of normal larvae (plutei) from unfertilized eggs of the sea urchin. *Am. J. Physiol.* 3:135–138.

Loehlin, J.C., and R.C. Nichols. 1976. *Heredity, Environment, and Personality: A Study of 850 Sets of Twins*. University of Texas Press, Austin.

Loi, P., G. Ptak, B. Barboni, J. Fulka Jr., P. Cappai, and M. Clinton. 2001. Genetic rescue of an endangered mammal by cross-species nuclear transfer using post-mortem somatic cells. *Nat. Biotechnol.* 19:962–964.

Longhurst, A.R. 1955. Evolution in the Notostraca. *Evolution* 9:84–86.

Loughry, W.J., G.M. Dwyer, and C.M. McDonald. 1998a. Behavioral interactions between juvenile nine-banded armadillos (*Dasypus novemcinctus*) in staged encounters. *Am. Midl. Nat.* 139:125–132.

Loughry, W.J., and C.M. McDonough. 1994. Scent discrimination by infant nine-banded armadillos. *J. Mammal.* 75:1033–1039.

Loughry, W.J., P.A. Prodöhl, C.M. McDonough, and J.C. Avise. 1998b. Polyembryony in armadillos. *Am. Sci.* 86:274–279.

Lowcock, L.A. 1989. Biogeography of hybrid complexes of *Ambystoma:* interpreting unisexual-bisexual genetic data in space and time. Pp. 180–208 in: *Evolution and Ecology of Unisexual Vertebrates,* R.M. Dawley and J.P. Bogart (eds.). New York State Museum, Albany.

Lowcock, L.A., and J.P. Bogart. 1989. Electrophoretic evidence for multiple origins of triploid forms in the *Ambystoma laterale-jeffersonianum* complex. *Can. J. Zool.* 67:350–356.

Lowcock, L.A., L.E. Licht, and J.P. Bogart. 1987. Nomenclature in hybrid complexes of *Ambystoma*: no case for the erection of hybrid "species." *Syst. Zool.* 36:328–336.

Lubinski, B.A., W.P. Davis, D.S. Taylor, and B.J. Turner. 1995. Outcrossing in a natural population of a self-fertilizing hermaphroditic fish. *J. Hered.* 86:469–473.

Lucas, M.D., R.E. Drew, P.A. Wheeler, P.A. Verrell, and G.H. Thorgaard. 2004. Behavioral differences among rainbow trout clonal lines. *Behav. Genet.* 34:355–365.

Lundmark, M. 2006. Polyploidization, hybridization and geographical parthenogenesis. *Trends Ecol. Evol.* 21:9.

Lynch, M. 1983. Ecological genetics of *Daphnia pulex*. *Evolution* 37:358–374.

Lynch, M. 1984a. The genetic structure of a cyclical parthenogen. *Evolution* 38:186–203.

Lynch, M. 1984b. Destabilizing hybridization, general purpose genotypes and geographic parthenogenesis. *Q. Rev. Biol.* 59:257–290.

Lynch, M., R. Bürger, D. Butcher, and W. Gabriel. 1993. The mutational meltdown in asexual populations. *J. Hered.* 84:339–344.

Lynch, M., J. Conery, and R. Burger. 1995. Mutational meltdowns in sexual populations. *Evolution* 49:1067–1080.

Lynch, M., and W. Gabriel. 1990. Mutation load and the survival of small populations. *Evolution* 44:1725–1737.

Lynch, M., B. Koskella, and S. Schaack. 2006. Mutation pressure and the evolution of organelle genomic architecture. *Science* 311:1727–1730.

MacCulloch, R.D., R.W. Murphy, L.A. Kupriyanova, and I.S. Darevsky. 1997. The Caucasian rock lizard *Lacerta rostombekovi:* a monoclonal parthenogenetic vertebrate. *Biochem. Syst. Ecol.* 25:33–37.

MacCulloch, R.D., R.W. Murphy, L.A. Kupriyanova, I.S. Darevsky, and F.D. Danielyan. 1995. Clonal variation in the parthenogenetic rock lizard *Lacerta armeniaca. Genome* 38:1057–1060.

MacGillivray, I., D.M. Campbell, and B. Thompson, eds. 1988. *Twinning and Twins.* Wiley, Chichester, UK.

Macgregor, H.C., and T.M. Uzzell Jr. 1964. Gynogenesis in salamanders related to *Ambystoma jeffersonianum. Science* 143:1043–1045.

Mackiewicz, M., A. Tatarenkov, A. Perry, J.R. Martin, D.F. Elder Jr., D.L. Bechler, and J.C. Avise. 2006a. Microsatellite documentation of male-mediated outcrossing between inbred laboratory strains of the self-fertilizing mangrove killifish (*Kryptolebias marmoratus*). *J. Hered.* 97:508–513.

Mackiewicz, M., A. Tatarenkov, D.S. Taylor, B.J. Turner, and J.C. Avise. 2006b. Extensive outcrossing and androdioecy in a vertebrate species that otherwise reproduces as a self-fertilizing hermaphrodite. *Proc. Natl. Acad. Sci. USA* 103:9924–9928.

Mackiewicz, M., A. Tatarenkov, B.J. Turner, and J.C. Avise. 2006c. A mixed-mating strategy in a hermaphroditic vertebrate. *Proc. R. Soc. Lond. B Biol. Sci.* 273:2449–2452.

Malysheva, D.N., O.N. Tokarskaya, V.G. Petrosyan, F.D. Danielyan, I.S. Darevsky, and A.P. Ryskov. 2007. Genomic variation in parthenogenetic lizard *Darevskia armeniaca:* evidence from DNA fingerprinting data. *J. Hered.* 98:173–178.

Mank, J.E., D.E.L. Promislow, and J.C. Avise. 2006. Evolution of alternative sex-determining mechanisms in Teleost fishes. *Biol. J. Linn. Soc.* 87:83–93.

Marchal, P. 1898. Le cycle évolutif de l'*Encyrtus fusicollis. Bull. Soc. Entomol. Fr.* 1898:109–111.

Markovic, M., and D. Trisovic. 1979. Monozygotic triplets with discordance for some traits. *Eur. J. Orthod.* 1:189–192.

Mark Welch, D.B., and M.S. Meselson. 2000. Evidence for the evolution of bdelloid rotifers without sexual reproduction or genetic exchange. *Science* 288:1211–1214.

Mark Welch, J.L, D.B. Mark Welch, and M. Meselson. 2004. Cytogenetic evidence for asexual evolution of bdelloid rotifers. *Proc. Natl. Acad. Sci. USA* 101:1618–1621.

Marshall, V.S., L.J. Wilton, and H.D.M. Moore. 1998. Parthenogenetic activation of marmoset (*Callithrix jacchus*) oocytes and the development of marmoset parthenogenones in vitro and in vivo. *Biol. Reprod.* 59:1491–1497.

Martens, K., G. Rossetti, and D.J. Horne. 2003. How ancient are ancient asexuals? *Proc. R. Soc. Lond. B Biol. Sci.* 270:723–729.

Martin, G.M. 2005. Epigenetic drift in aging identical twins. *Proc. Natl. Acad. Sci. USA* 102:10413–10414.

Martins, M.J., M.J. Collares-Pereira, I.G. Cowx, and M.M. Coelho. 1998. Diploids v. triploids of *Rutilus alburnoides:* spatial segregation and morphological differences. *J. Fish Biol.* 52:817–828.

Maslin, T.P. 1968. Taxonomic problems in parthenogenetic vertebrates. *Syst. Zool.* 17:219–231.

Massaro, E.J., J.C. Massaro, and R.W. Harrington Jr. 1975. Biochemical comparison of genetically different homozygous clones (isogenic, uniparental lines) of the self-fertilizing fish *Rivulus marmoratus* Poey. Pp. 439–453 in: *Isozymes,* vol. 3, C.L. Markert (ed.). Academic Press, New York.

Mateos, M., O.I. Sanjur, and R.C. Vrijenhoek. 2002. Historical biogeography of the livebearing fish genus *Poeciliopsis* (Poeciliidae: Cyprinodontiformes). *Evolution* 56:972–984.

Mateos, M., and R.C. Vrijenhoek. 2002. Ancient versus reticulate origin of a hemiclonal lineage. *Evolution* 56:985–992.

Mateos, M., and R.C. Vrijenhoek. 2005. Independent origins of allotriploidy in the fish genus *Poeciliopsis. J. Hered.* 96:32–39.

Maynard Smith, J. 1971. The origin and maintenance of sex. Pp. 163–175 in: *Group Selection,* G.C. Williams (ed.). Aldine-Atherton, Chicago.

Maynard Smith, J. 1978. *The Evolution of Sex.* Cambridge University Press, Cambridge, UK.

Maynard Smith, J. 1986. Contemplating life without sex. *Nature* 324:300–301.

Maynard Smith, J. 1992. Age and the unisexual lineage. *Nature* 356:661–662.

Maynard Smith, J., and J. Haigh. 1974. The hitch-hiking effect of a favourable gene. *Genet. Res.* 23:23–35.

McDonough, C.M., and W.J. Loughry. 1997. Patterns of mortality in a population of nine-banded armadillos (*Dasypus novemcinctus*). *Am. Midl. Nat.* 138:299–305.

McGloughlin, M.N., and J.I. Burke. 2000. *Biotechnology: Present Position and Future Development.* Teagasc Publishers, Dublin, Ireland.

McGregor, H.C., and T. Uzzell. 1964. Gynogenesis in salamanders related to *Ambystoma jeffersonianum. Science* 143:1043–1045.

McKay, F.E. 1971. Behavioral aspects of population dynamics in unisexual-bisexual *Poeciliopsis* (Pisces: Poeciliidae). *Ecology* 52:778–790.

McKone, M.J., and S.L. Halpern. 2003. The evolution of androgenesis. *Am. Nat.* 161:641–656.

McVean, G.A.T. 2001. What do patterns of genetic variability reveal about mitochondrial recombination? *Heredity* 87:613–620.

Mead, R.A. 1989. The physiology and evolution of delayed implantation in carnivores. Pp. 437–464 in: *Carnivore Behavior, Ecology, and Evolution,* J.L. Gittleman (ed.). Cornell University Press, Ithaca, NY.

Medawar, P. 1957. *The Uniqueness of the Individual.* Dover, New York..

Medawar, P. 1958. The Croonian lecture. The homograft reaction. *Proc. R. Soc. Lond. B Biol. Sci.* 149:145–166.

Meng, L., J.J. Ely, R.L. Stouffer, and D.P. Wolf. 1997. Rhesus monkeys produced by nuclear transfer. *Biol. Reprod.* 57:454–459.

Mesbah, S.F., M. Kafi, H. Hili, and M.H. Nasr-Esfahani. 2004. Spontaneous parthenogenesis and development of camel (*Camelus dromedarius*) oocytes. *Vet. Rec.* 155:498–500.

Meselson, M., and F.W. Stahl. 1958. The replication of DNA in *Escherichia coli. Proc. Natl. Acad. Sci. USA* 44:671–682.

Michod, R.M., and B.R. Levin, eds. 1988. *The Evolution of Sex.* Sinauer, Sunderland, MA.

Miller, R.R., and R.J. Schultz. 1959. All-female strains of the teleost fishes of the genus *Poeciliopsis. Science* 130:1656–1657.

Minton, S.A. 1954. Salamanders of the *Ambystoma jeffersonianum* complex in Indiana. *Herpetologica* 10:173–179.

Mitalipov, S.M., K.L. White, V.R. Farrar, J. Morrey, and W.A. Reed. 1999. Development of nuclear transfer and parthenogenetic rabbit embryos activated with inositol 1,4,5-triphosphate. *Biol. Reprod.* 60:821–827.

Modlinsky, J.A. 1978. Transfer of embryonic nuclei to fertilized mouse eggs and development of tetraploid blastocysts. *Nature* 273:466–467.

Monaco, P.J., E.M. Rasch, and P.R. Musich. 1984. Apomictic reproduction in the Amazon molly, *Poecilia formosa,* and its triploid hybrids. Pp. 311–328 in: *Evolutionary Genetics of Fishes,* B.J. Turner (ed.). Plenum Press, New York.

Moore, W.S. 1975. Stability of unisexual-bisexual populations of *Poeciliopsis* (Pisces: Poeciliidae). *Ecology* 56:791–808.

Moore, W.S. 1976. Components of fitness in the unisexual fish *Poeciliopsis monacha-occidentalis. Evolution* 30:564–578.

Moore, W.S. 1977. A histocompatibility analysis of inheritance in the unisexual fish *Poeciliopsis 2 monacha-lucida. Copeia* 1977:213–223.

Moore, W.S., and F.E. McKay. 1971. Coexistence in unisexual-bisexual species complexes of *Poeciliopsis* (Pisces: Poeciliidae). *Ecology* 52:791–799.

Moore, W.S., R.R. Miller, and R.J. Schultz. 1970. Distribution, adaptation and probable origin of an all-female form of *Poeciliopsis* (Pisces: Poecilidae) in northwestern Mexico. *Evolution* 24:789–795.

Moran, N.A. 2007. Symbiosis as an adaptive process and source of phenotypic complexity. Pp. 165–181 in: *In the Light of Evolution I: Adaptation and Complex Design,* J.C. Avise and F.J. Ayala (eds.). National Academy Press, Washington, DC.

Morishima, K., S. Horie, E. Yamaha, and K. Arai. 2002. A cryptic clonal line of the loach *Misgurnus anguillicaudatus* (Teleostei: Cobitidae) evidenced by induced gynogenesis, interspecific hybridization, microsatellite genotyping and multilocus DNA fingerprinting. *Zool. Sci.* 19:565–575.

Morishima, K., Y. Nakamura-Shiokawa, E. Bando, Y.-J. Li, A. Boron, M.R. Khan, and K. Arai. 2007. Cryptic clonal lineages and genetic diversity in the loach *Misgurnus anguillicaudatus* (Teleostei: Cobitidae) inferred from nuclear and mitochondrial DNA analyses. *Genetica* 132:159–171.

Morishima, K., K. Oshima, S. Horie, T. Fujimoto, E. Yamaha, and K. Arai. 2004. Clonal diploid sperm of the diploid-triploid mosaic loach, *Misgurnus anguillicaudatus* (Teleostei: Cobitidae). *J. Exp. Zool.* 301A:502–511.

Morison, I.M., J.P. Ramsey, and H.G. Spencer. 2005. A census of mammalian imprinting. *Trends Genet.* 21:457–465.

Morito, Y., Y. Terada, S. Nakamura, J. Morita, T. Yoshimoto, T. Murakami, N. Yaegashi, and K. Okamura. 2005. Dynamics of microtubules and positioning of female pronucleus during bovine parthenogenesis. *Biol. Reprod.* 73:935–941.

Moritz, C. 1991. The origin and evolution of parthenogenesis in *Heteronotia binoei* (Gekkonidae): evidence for recent and localized origins of widespread clones. *Genetics* 129:211–219.

Moritz, C. 1993. The origin and evolution of parthenogenesis in the *Heteronotia binoei* complex: synthesis. *Genetica* 90:269–280.

Moritz, C., M. Adams, S. Donnellan, and P. Baverstock. 1990. The origin and evolution of parthenogenesis in *Heteronotia binoei* (Gekkonidae): genetic diversity among bisexual populations. *Copeia* 1990:333–348.

Moritz, C., W.M. Brown, L.D. Densmore, J.W. Wright, D. Vyas, S. Donnellan, M. Adams, and P. Baverstock. 1989a. Genetic diversity and the dynamics of hybrid parthenogenesis in *Cnemidophorus* (Teiidae) and *Heteronotia* (Gekkonidae). Pp. 87–112 in: *Evolution and Ecology of Unisexual Vertebrates,* R.M. Dawley and J.P. Bogart (eds.). New York State Museum, Albany.

Moritz, C., T.J. Case, D.T. Bolger, and S. Donnellan. 1993. Genetic diversity and the history of Pacific island house geckos. *Biol. J. Linn. Soc.* 48:113–133.

Moritz, C., S. Donnellan, M. Adams, and P.R. Baverstock. 1989b. The origin and evolution of parthenogenesis in *Heteronotia binoei* (Gekkonidae): extensive genotypic diversity among parthenogens. *Evolution* 43:994–1003.

Moritz, C., and A. Heideman. 1993. The origin and evolution of parthenogenesis in *Heteronotia binoei* (Gekkonidae): reciprocal origins and diverse mitochondrial DNA in western populations. *Syst. Biol.* 42:293–306.

Moritz, C., H. McCallum, S. Donnellan, and J. Roberts. 1991. Parasite loads in parthenogenetic and sexual lizards (*Heteronotia binoei*): support for the red queen hypothesis. *Proc. R. Soc. Lond. B Biol. Sci.* 244:145–149.

Moritz, C., T. Uzzell, S. Spolsky, H. Hotz, I. Darevsky, L. Kupriyanova, and F. Danielyan. 1992a. The maternal ancestry and approximate age of parthenogenetic species of Caucasian rock lizards (*Lacerta:* Lacertidae). *Genetica* 87:53–62.

Moritz, C., J.W. Wright, and W.M. Brown. 1989c. Mitochondrial-DNA analyses and the origin and relative age of parthenogenetic lizards (*Cnemidophorus*). III. *C. velox* and *C. exsanguis. Evolution* 43:958–968.

Moritz, C., J.W. Wright, and W.M. Brown. 1992b. Mitochondrial DNA analyses and the origin and relative age of parthenogenetic *Cnemidophorus:* phylogenetic constraints on hybrid origins. *Evolution* 46:184–192.

Mortensen, T.H. 1921. *Studies of the Development and Larval Forms of Echinoderms.* G.E.C. Gad, Copenhagen.

Muller, H.J. 1932. Some genetic aspects of sex. *Am. Nat.* 66:118–138.

Muller, H.J. 1964. The relation of mutation to mutational advance. *Mutat. Res.* 1:2–9.

Murphy, R.W., I.S. Darevsky, R.D. MacCulloch, J. Fu, L.A. Kupriyanova, D.E. Upton, and F. Danielyan. 1997. Old age, multiple formations or genetic plasticity? Clonal diversity in the uniparental Caucasian rock lizard, *Lacerta dahli. Genetica* 101:125–130.

Murphy, R.W., J.Z. Fu, R.D. MacCulloch, I.S. Darevsky, and I.S. Kupriyanova. 2000. A fine line between sex and unisexuality: the phylogenetic constraints on parthenogenesis in lacertid lizards. *Zool. J. Linn. Soc.* 130:527–549.

Murphy, W.J., J.E. Thomerson, and G.E. Collier. 1999. Phylogeny of the neotropical killifish family Rivulidae (Cyprinodontiformes, Aplocheiloidei) inferred from mitochondrial DNA sequences. *Mol. Phylogenet. Evol.* 13:289–301.

Nace, G.W., C.M. Richards, and J.H. Asher. 1970. Parthenogenesis and genetic variability. I. Linkage and inbreeding estimations in the frog, *Rana pipiens. Genetics* 66:349–368.

Nagylaki, T. 1976. A model for the evolution of self-fertilization and vegetative reproduction. *J. Theor. Biol.* 58:55–58.

Nam, Y.K., Y.S. Cho, and D.S. Kim. 2000. Isogenic transgenic homozygous fish induced by artificial parthenogenesis. *Transgenic Res.* 9:463–469.

Nanda, I., I. Schlupp, D.K. Lamatsch, K.P. Lampert, M. Schmid, and M. Schartl. 2007. Stable inheritance of host species-derived microchromosomes in the gynogenetic fish, *Poecilia formosa. Genetics* 177:917–926.

Naruse, K., K. Ijiri, A. Shima, and E. Egami. 1985. The production of cloned fish in the medaka (*Oryzias latipes*). *J. Exp. Zool.* 236:335–341.

Neaves, W.B. 1971. Tetraploidy in a hybrid lizard of the genus *Cnemidophorus* (Teiidae). *Breviora* 381:1–25.

Negovetic, S., B.R. Anholt, R.D. Semlitsch, and H.-U. Reyer. 2000. Specific responses of sexual and hybridogenetic European waterfrog tadpoles to temperature. *Ecology* 82:766–774.

Newby, E. 1966. *Slowly Down the Ganges.* Hodder and Stoughton, London.

Newman, H.H. 1913. The natural history of the nine-banded armadillo of Texas. *Am. Nat.* 47:513–539.

Newman, H.H. 1923. *The Physiology of Twinning.* University of Chicago Press, Chicago.

Newman, H.H., and J.T. Patterson. 1909. A case of normal identical quadruplets in the nine-banded armadillo, and its bearings on the problems of identical twins and of sex-determination. *Biol. Bull.* 17:181–187.

Newman, H.H., and J.T. Patterson. 1910. The development of the nine-banded armadillo from the primitive strek stage to birth: with special reference to the question of specific polyembryony. *J. Morphol.* 21:359–423.

Nichols, K.M., W.P. Young, R.G. Danzmann, B.D. Robison, C. Rexroad, M. Noakes, R.B. Phillips, P. Bentzen, I. Spies, K. Knudsen, et al. 2003. An updated genetic linkage map for rainbow trout (*Oncorhynchus mykiss*). *Anim. Genet.* 34:102–115.

Niemeitz, A., R. Kreutzfeldt, M. Schartl, J. Parzefall, and I. Schlupp. 2002. Male mating behavior of a molly, *Poecilia latipunctata:* a third host for the sperm-dependent Amazon molly, *Poecilia formosa. Acta Ethol.* 5:45–49.

Ode, P.J., and M.R. Strand. 1995. Progeny and sex allocation decisions of the polyembryonic wasp *Copidosoma floridanum. J. Anim. Ecol.* 64:213–224.

Ogoh, K., and Y. Ohmiya. 2007. Concerted evolution of duplicated control regions within an ostracod mitochondrial genome. *Mol. Biol. Evol.* 24:74–78.

Ohta, T. 2000. Evolution of gene families. *Gene* 259:45–52.

Olsen, M.W. 1962. Polyembryony in unfertilized turkey eggs. *J. Hered.* 53:125–129.

Olsen, M.W. 1974. Frequency and cytological aspects of diploid parthenogenesis in turkey eggs. *Theor. Appl. Genet.* 44:216–221.

Owusu-Frimpong, M., and J.A. Hargreaves. 2000. Incidence of conjoined twins in tilapia after thermal shock induction of polyploidy. *Aquac. Res.* 31:421–426.

Ozil, J.P. 1990. The parthenogenetic development of rabbit oocytes after repetitive pulsatile electrical stimulation. *Development* 109:117–127.

Pannell, J.R. 1997. The maintenance of gynodioecy and androdioecy in a metapopulation. *Evolution* 51:10–20.

Pannell, J.R. 2002. The evolution and maintenance of androdioecy. *Annu. Rev. Ecol. Syst.* 33:397–425.

Papaioannou, V.E., J. Mkandawire, and J.D. Biggers. 1989. Development and phenotypic variability of genetically identical half mouse embryos. *Development* 106:817–827.

Parker, E.D. 1979. Ecological implications of clonal diversity in parthenogenetic morphospecies. *Am. Zool.* 19:753–762.

Parker, E.D., and R.K. Selander. 1976. The organization of genetic diversity in the parthenogenetic lizard *Cnemidophorus neomexicanus* (Sauria: Teiidae). *Genetics* 84:791–805.

Parker, E.D., and R.K. Selander. 1984. Low clonal diversity in the parthenogenetic lizard *Cnemidophorus neomexicanus* (Sauria, Teiidae). *Herpetologica* 40:245–252.

Parker, E.D., R.K. Selander, R.O. Hudson, and L.J. Lester. 1977. Genetic diversity in parthenogenetic cockroaches. *Evolution* 31:836–842.

Parker, E.D., Jr., J.M. Walker, and M.A. Paulissen. 1989. Clonal diversity in *Cnemidophorus:* ecological and morphological consequences. Pp. 72–86 in: *Evolution and Ecology of Unisexual Vertebrates,* R.M. Dawley and J.P. Bogart (eds.). New York State Museum, Albany.

Parker, G.A. 1985. Models of parent-offspring conflict. 5. Effects of the behavior of two parents. *Anim. Behav.* 33:519–533.

Parsons, J.E., and G.H. Thorgaard. 1984. Induced androgenesis in rainbow trout. *J. Exp. Zool.* 231:407–412.

Parsons, J.E., and G.H. Thorgaard. 1985. Production of androgenetic diploid rainbow trout. *J. Hered.* 76:177–181.

Paschos, I., L. Natsis, C. Nathanailides, I. Kagalou, and E. Kolettas. 2001. Induction of gynogenesis and androgenesis in goldfish *Carassius auratus* (var. *oranda*). *Reprod. Domest. Anim.* 36:195–198.

Pattee, O.H., W.G. Mattox, and W.S. Seegar. 1984. Twin embryos in a peregrine falcon egg. *Condor* 86:352–353.

Patterson, J.T. 1913. Polyembryonic development in *Tatusia novemcincta. J. Morphol.* 24:559–683.

Patterson, J.T. 1927. Polyembryony in animals. *Q. Rev. Biol.* 2:399–426.

Pauly, D. 2004. *Darwin's Fishes.* Cambridge University Press, Cambridge, UK.

Pearse, D.E., F.J. Janzen, and J.C. Avise. 2001. Genetic markers substantiate long-term storage and utilization of sperm by female painted turtles. *Heredity* 86: 378–384.

Pellegrino, K.C.M., M.T. Rodrigues, and Y. Yonenaga-Yassuda. 2003. Triploid karyotype of *Leposoma percarinatum* (Squamata, Gymnophthalmidae). *J. Herpetol.* 37:197–199.

Petren, K., D.T. Bolger, and T.J. Case. 1993. Mechanisms in the competitive success of an invading sexual gecko over and asexual native. *Science* 259:354–358.

Polejaeva, I.A., S.H. Chen, T.D. Vaught, R.L. Page, J. Mullins, S. Ball, Y. Dai, J. Boone, S. Walker, D.L. Ayares, et al. 2000. Cloned pigs produced by nuclear transfer from adult somatic cells. *Nature* 407:86–90.

Prather, R.S., M.M. Sims, and N.L. First. 1989. Nuclear transplantation in early pig embryos. *Biol. Reprod.* 41:414–418.

Prodöhl, P.A., W.J. Loughry, C.M. McDonough, W.S. Nelson, and J.C. Avise. 1996. Molecular documentation of polyembryony and the micro-spatial dispersion of clonal sibships in the nine-banded armadillo. *Proc. R. Soc. Lond. B Biol. Sci.* 263:1643–1649.

Prodöhl, P.A., W.J. Loughry, C.M. McDonough, W.S. Nelson, E.A. Thompson, and J.C. Avise. 1998. Genetic maternity and paternity in a local population of armadillos assessed by microsatellite DNA markers and field data. *Am. Nat.* 151:7–19.

Purdom, C.E. 1969. Radiation-induced gynogenesis and androgenesis in fish. *Heredity* 24:431–444.

Quattro, J.M., J.C. Avise, and R.C. Vrijenhoek. 1991. Molecular evidence for multiple origins of hybridogenetic fish clones (Poeciliidae: *Poeciliopsis*). *Genetics* 127:391–398.

Quattro, J.M., J.C. Avise, and R.C. Vrijenhoek. 1992a. An ancient clonal lineage in the fish genus *Poeciliopsis* (Atheriniformes: Poeciliidae). *Proc. Natl. Acad. Sci. USA* 89:348–352.

Quattro, J.M., J.C. Avise, and R.C. Vrijenhoek. 1992b. Mode of origin and sources of genotypic diversity in triploid fish clones (*Poeciliopsis:* Poeciliidae). *Genetics* 130:621–628.

Radtkey, R.R., B. Becker, R.D. Miller, R. Riblet, and T.J. Case. 1996. Variation and evolution of class I Mhc in sexual and parthenogenetic geckos. *Proc. R. Soc. Lond. B Biol. Sci.* 263:1023–1032.

Radtkey, R.R., S.C. Donnellan, R.N. Fisher, C. Moritz, K.A. Hanley, and T.J. Case. 1995. When species collide: the origin and spread of an asexual species of gecko. *Proc. R. Soc. Lond. B Biol. Sci.* 259:145–152.

Ragghianti, M., S. Bucci, S. Marracci, C. Casola, G. Mancino, H. Hotz, G.D. Guex, J. Plötner, and T. Uzzell. 2007. Gametogenesis of intergroup hybrids of hemiclonal frogs. *Genet. Res.* 89:39–45.

Raikova, E.V. 1980. Morphology, ultrastructure and development of the parasitic larva and its surrounding trophamnion of *Polypodium hydriforme* Coelenterata. *Cell Tissue Res.* 206:487–500.

Ralls, K., and J. Ballou. 1983. Extinction: lessons from zoos. Pp. 164–184 in: *Genetics and Conservation,* C.M. Shonewald-Cox, S.M. Chambers, B. MacBryde, and L. Thomas (eds.). Benjamin/Cummings, Menlo Park, CA.

Randall, J.E. 2005. *Reef and Shore Fishes of the South Pacific.* University of Hawaii Press, Honolulu.

Rasch, E.M., and P.J. Balsano. 1989. Trihybrids related to the unisexual molly fish, *Poecilia formosa.* Pp. 252–267 in: *Evolution and Ecology of Unisexual Vertebrates,* R.M. Dawley and J.P. Bogart (eds.). New York State Museum, Albany.

Rasch, E.M., P.J. Monaco, and J.S. Balsano. 1982. Cytophotometric and autoradiographic evidence for functional apomixes in a gynogenetic fish, *Poecilia* and its related triploid unisexuals. *Histochemistry* 73:515–533.

Rasch, E.M., L.M. Prehn, and R.W. Rasch. 1970. Cytogenetic studies of *Poecilia* (Pisces). II. Triploidy and DNA levels in naturally occurring populations associated with the gynogenetic teleost, *Poecilia formosa* (Girard). *Chromosoma* 31:18–40.

Raven, P.H. 1979. A survey of reproductive biology in Onagraceae. *N.Z. J. Bot.* 17:575–593.

Reddy, U.M., A.M. Branum, and M.A. Klebanoff. 2005. Relationship of maternal body mass index and height to twinning. *Obstet. Gynecol.* 105:593–597.

Reed, C.G. 1991. Bryozoa. Pp. 85–245 in: *Reproduction of Marine Invertebrates. VI. Echnioderms and Lophophorates,* A.C. Giese and J.S. Pearse (eds.). Boxwood Press, Pacific Grove, CA.

Reeder, T., H.C. Dessauer, and C.J. Cole. 2002. Phylogenetic relationships of whiptail lizards of the genus *Cnemidophorus* (Squamata, Teiidae): a test of monophyly, reevaluation of karyotypic evolution, and review of hybrid origins. *Am. Mus. Novit.* 3365:1–61.

Reeve, H.K., D.F. Westneat, W.A. Noon, P.W. Sherman, and C.F. Aquadro. 1990. DNA "fingerprinting" reveals high levels of inbreeding in colonies of the eusocial naked mole-rat. *Proc. Natl. Acad. Sci. USA* 87:2496–2500.

Renfree, M.B. 1978. Embryonic diapause in mammals: a developmental strategy. Pp. 1–46 in: *Dormancy and Developmental Arrest,* M.E. Clutter (ed.). Academic Press, New York.

Riley, W.A. 1907. Polyembryony and sex determination. *Science* 25:106–107.

Rist, L., R.D. Semlitsch, H. Hotz, and H.-U. Reyer. 1997. Feeding behavior, food consumption, and growth efficiency of hemiclonal and parental tadpoles of the *Rana esculenta* complex. *Funct. Ecol.* 11:735–742.

Ritchie, S.A., and R.P. Davis. 1986. Evidence for embryonic diapause in *Rivulus marmoratus:* laboratory and field observations. *J. Am. Killifish Assoc.* 19:103–108.

Robertson, D.R. 1972. Social control of sex-reversal in a coral-reef fish. *Science* 117:1007–1009.

Robinette, W.L., N.V. Hancock, and D.A. Jones. 1977. *The Oak Creek Mule Deer Herd in Utah.* Utah State Division of Wildlife Resources, Salt Lake City.

Robison, B.D., P.A. Wheeler, and G.H. Thorgaard. 1999. Variation in development rate among clonal lines of rainbow trout (*Oncorhynchus mykiss*). *Aquaculture* 173:131–141.

Rocha, C.F.D., H.G. Bergallo, and D. Peccinini-Seale. 1997. Evidence of an unisexual population for the Brazilian whiptail lizard genus *Cnemidophorus* (Teiidae), with description of a new species. *Herpetologica* 53:374–382.

Rokas, A., E. Ladoukakis, and E. Zouros. 2003. Animal mitochondrial DNA recombination revisited. *Trends Ecol. Evol.* 18:411–417.

Rougier, N., and Z. Werb. 2001. Minireview: parthenogenesis in mammals. *Mol. Reprod. Dev.* 59:468–474.

Russell, W.C., J.S. Brinks, and G.V. Richardson. 1984. Changes in genetic variances with increased inbreeding of beef cattle. *J. Hered.* 75:8–10.

Ryland, J.D. 1996. Polyembryony "paradox": the case of cyclostomate Bryozoa. *Trends Ecol. Evol.* 11:26.

Ryland, J.S. 1970. *Bryozoans.* Hutchinson, London.

Sakakura, Y., and D.L.G. Noakes. 2000. Age, growth, and sexual development in the self-fertilizing hermaphroditic fish *Rivulus marmoratus. Environ. Biol. Fishes* 59:309–317.

Sakakura, Y., K. Soyano, D.L.G. Noakes, and A. Hagiwara. 2006. Gonadal morphology in the self-fertilizing mangrove killifish, *Kryptolebias marmoratus. Ichthyol. Res.* 53:427–430.

Sarder, M.R.I., D.J. Penman, J.M. Myers, and B.J. McAndrew. 1999. Production and propagation of fully inbred clonal lines in the Nile tilapia (*Oreochromis niloticus* L.). *J. Exp. Zool.* 284:675–685.

Sato, A., Y. Satta, F. Figueroa, W.E. Mayer, Z. Zaleska-Rutczynska, S. Toyosawa, J. Travis, and J. Klein. 2002. Persistence of *Mhc* heterozygosity in homozygous clonal killifish, *Rivulus marmoratus:* implications for the origin of hermaphroditism. *Genetics* 162:1791–1803.

Satoh, M., and T. Kuroiwa. 1991. Organization of multiple nucleoids and DNA molecules in mitochondria of a human cell. *Exp. Cell Res.* 196:137–140.

Schartl, M., I. Nanda, I. Schlupp, B. Wilde, J.T. Epplen, M. Schmid, and J. Parzefall. 1995a. Incorporation of subgenomic amounts of DNA as compensation for mutational load in a gynogenetic fish. *Nature* 373:68–71.

Schartl, M., I. Schlupp, A. Schartl, M.K. Meyer, I. Nanda, M. Schmid, J.T. Epplen, and J. Parzefall. 1991. On the stability of dispensable constituents of the eukaryotic genome: stability of coding sequences versus truly hypervariable sequences in a clonal vertebrate, the Amazon molly, *Poecilia formosa. Proc. Natl. Acad. Sci. USA* 88:8759–8763.

Schartl, M., B. Wilde, I. Schlupp, and J. Parzefall. 1995b. Evolutionary origin of a parthenoform, the Amazon molly *Poecilia formosa,* on the basis of a molecular phylogeny. *Evolution* 49:827–835.

Schatten, G., R. Prather, and I. Wilmut. 2003. Cloning claim is science fiction, not science. *Science* 299:344.

Scheerer, P.D., G.H. Thorgaard, and F.W. Allendorf. 1991. Genetic analysis of androgenetic rainbow trout. *J. Exp. Zool.* 260:382–390.

Scheerer, P.D., G.H. Thorgaard, F.W. Allendorf, and K.I. Knudsen. 1986. Androgenetic rainbow trout produced from inbred and outbred sperm sources show similar survival. *Aquaculture* 57:289–298.

Schemske, D.W., and R. Lande. 1985. The evolution of self-fertilization and inbreeding depression in plants. II. Empirical observations. *Evolution* 39:41–52.

Schenck, R.A., and R.C. Vrijenhoek. 1986. Spatial and temporal factors affecting coexistence among sexual and clonal forms of *Poeciliopsis. Evolution* 40:1060–1070.

Schenck, R.A., and R.C. Vrijenhoek. 1989. Coexistence among sexual and asexual forms of *Poeciliopsis:* foraging behavior and microhabitat selection. Pp. 39–48 in: *Evolution and Ecology of Unisexual Vertebrates,* R.M. Dawley and J.P. Bogart (eds.). New York State Museum, Albany.

Schlosser, I.J., M.R. Doeringsfeld, J.F. Endler, and L.F. Arzayus. 1997. Niche relationships of clonal and sexual fish in a heterogeneous landscape. *Ecology* 79:953–968.

Schlupp, I. 2005. The evolutionary ecology of gynogenesis. *Annu. Rev. Ecol. Evol. Syst.* 36:399–417.

Schlupp, I., C. Marler, and M.J. Ryan. 1994. Benefit to male sailfin mollies of mating with heterospecific females. *Science* 263:373–374.

Schlupp, I., J. Parzefall, and M. Schartl. 1991. Male mate choice in mixed bisexual/unisexual breeding complexes of *Poecilia* (Teleostei: Poeciliidae). *Ethology* 88:215–222.

Schlupp, I., J. Parzefall, and M. Schartl. 2002. Biogeography of the unisexual Amazon molly, *Poecilia formosa*. *J. Biogeogr.* 29:1–6.

Schories, S., K.P. Lampert, D.K. Lamatsch, F.J. García de León, and M. Schartl. 2007. Analysis of a possible independent origin of triploid *P. formosa* outside of the Río Purificación river system. *Front. Zool.* 4:13.

Schuett, G.W., P.J. Fernandez, D. Chiszar, and M.H. Smith. 1998. Fatherless sons: a new type of parthenogenesis in snakes. *Fauna* 1:19–25.

Schuett, G.W., P.J. Fernandez, W.F. Gergits, N.J. Casna, D. Chiszar, H.M. Smith, J.B. Mitton, S.P. Mackessy, R.A. Odum, and M.J. Denlong. 1997. Production of offspring in the absence of males: evidence for facultative parthenogenesis in bisexual snakes. *Herpetol. Nat. Hist.* 5:1–10.

Schultz, R.J. 1961. Reproductive mechanism of unisexual and bisexual strains of the viviparous fish *Poeciliopsis*. *Evolution* 15:302–325.

Schultz, R.J. 1967. Gynogenesis and triploidy in the viviparous fish *Poeciliopsis*. *Science* 157:1564–1567.

Schultz, R.J. 1969. Hybridization, unisexuality and polyploidy in the teleost *Poeciliopsis* (Poeciliidae) and other vertebrates. *Am. Nat.* 103:605–619.

Schultz, R.J. 1971. Special adaptive problems associated with unisexual fishes. *Am. Zool.* 11:351–360.

Schultz, R.J. 1973. Unisexual fish: laboratory synthesis of a "species." *Science* 179:180–181.

Schultz, R.J. 1977. Evolution and ecology of unisexual fishes. *Evol. Biol.* 10:277–331.

Schultz, R.J. 1982. Competition and adaptation among diploid and polyploid clones of unisexual fishes. Pp. 103–119 in: *Evolution and Genetics of Life Histories*, H. Dingle and J.P. Hegmann (eds.). Springer, Berlin.

Schultz, R.J., and K.D. Kallman. 1968. Triploid hybrids between the all-female teleost *Poecilia formosa* and *Poecilia sphenops*. *Nature* 219:280–282.

Seike, N., K. Utaka, and H. Kanagawa. 1990. Production and development of calves from sexed-bisected bovine embryos. *Jpn. J. Vet. Res.* 38:1–9.

Selander, R.K., and R.O. Hudson. 1976. Animal population structure under close inbreeding: the land snail *Rumina* in southern France. *Am. Nat.* 110:695–718.

Selander, R.K., and D.W. Kaufman. 1975. Genetic population structure and breeding systems. *Isozymes* 4:27–48.

Semlitsch, R.D. 1993. Asymmetric competition in mixed populations of tadpoles of the hybridogenetic *Rana esculenta* complex. *Evolution* 47:510–519.

Semlitsch, R.D., S. Schmiedehausen, H. Hotz, and P. Beerli. 1996. Genetic compatibility between sexual and clonal genomes in local populations of the hybridogenetic *Rana esculenta* complex. *Evol. Ecol.* 10:531–543.

Sengoku, K., N. Takuma, T. Miyamato, T. Yamauchi, and M. Ishikawa. 2004. Nuclear dynamics of parthenogenesis of human oocytes: effect of oocyte aging in vitro. *Gynecol. Obstet. Invest.* 58:155–159.

Shao, C., L. Deng, O. Henegariu, L. Liang, N. Raikwar, A. Sahota, P.J. Stambrook, and J.A. Tischfield. 1999. Mitotic recombination produces the majority of recessive fibroblast variants in heterozygous mice. *Proc. Natl. Acad. Sci. USA* 96:9230–9235.

Shin, T., D. Kraemer, J. Pryor, L. Liu, J. Rugila, L. Howe, S. Buck, K. Murphy, L. Lyons, and M. Westhusin. 2002. A cat cloned by nuclear transplantation. *Nature* 415:859.

Shor, E.N., R.H. Rosenblatt, and J.D. Isaacs. 1987. Carl Leavitt Hubbs. *Biogr. Mem. Nat1. Acad. Sci.* 56:215–249.

Silver, L.M. 1995. *Mouse Genetics: Concepts and Applications.* Oxford University Press, New York.

Simon, J.-C., F. Delmotte, C. Rispe, and T. Crease. 2003. Phylogenetic relationships between parthenogens and their sexual relatives: the possible routes to parthenogenesis in animals. *Biol. J. Linn. Soc.* 79:151–163.

Sinclair, E.A., R. Scholl, R.L. Bezy, K.A. Crandall, and J.W. Sites Jr. 2006. Isolation and characterization of di- and tetranucleotide microsatellite loci in the yellow-spotted night lizard *Lepidophyma flavimaculatum* (Squamata: Xantusiidae). *Mol. Ecol. Notes* 6:233–236.

Sites, J.W., Jr., D.M. Peccinini-Seale, C. Moritz, J.W. Wright, and W.M. Brown. 1990. The evolutionary history of parthenogenetic *Cnemidophorus lemniscatus* (Sauria, Teiidae). I. Evidence for a hybrid origin. *Evolution* 44:906–921.

Smith, H.M., and E.D. Brodie Jr. 1982. *Reptiles of North America.* Golden Press, New York.

Snell, G.D. 1957. The homograft reaction. *Annu. Rev. Microbiol.* 11:439–458.

Spolsky, C.M., C.A. Phillips, and T. Uzzell. 1992. Antiquity of clonal salamander lineages revealed by mitochondrial DNA. *Nature* 356:706–708.

Spolsky, C.M., and T. Uzzell. 1984. Natural interspecies transfer of mitochondrial DNA in amphibians. *Proc. Natl. Acad. Sci. USA* 81:5802–5805.

Spolsky, C.M., and T. Uzzell. 1986. Evolutionary history of the hybridogenetic hybrid frog *Rana esculenta* as deduced from mtDNA analyses. *Mol. Biol. Evol.* 3:44–56.

Staats, J. 1966. The laboratory mouse. Pp. 1–9 in: *Biology of the Laboratory Mouse,* E.L. Green (ed.). McGraw-Hill, New York.

Stamps, J.A., R. Metcalf, and V.V. Krishnan. 1978. Genetic analysis of parent-offspring conflict. *Behav. Ecol. Sociobiol.* 3:369–392.

Stanley, J.G. 1976. Production of hybrid, androgenetic, and gynogenetic grass carp and carp. *Trans. Am. Fish. Soc.* 105:10–16.

Stanley, J.G., C.J. Biggers, and D.E. Schultz. 1976. Isozymes in androgenetic and gynogenetic white amur, gynogenetic carp, and carp-amur hybrids. *J. Hered.* 67:129–134.

Stanley, J.G., and K.E. Sneed. 1974. Artificial gynogenesis and its application in genetics and selective breeding of fishes. Pp. 527–536 in: *The Early Life History of Fish,* J.H.S. Blaxter (ed.). Springer-Verlag, New York.

Stebbins, G.L. 1957. Self fertilization and population variability in higher plants. *Am. Nat.* 91:337–354.

Steinman, G. 1998. Spontaneous monozygotic quadruplet pregnancy: an obstetric rarity. *Obstet. Gynecol.* 91:866.

Stenseth, N.C. and L.R. Kirkendall. 1985. On the evolution of pseudogamy. *Evolution* 39:294–307.

Stöck, M., D.K. Lamatsch, C. Steinlein, J.T. Epplen, W.R. Grosse, R. Hock, T. Klapperstück, K.P. Lampert, U. Scheer, M. Schmid, et al. 2002. A bisexually reproducing all-triploid vertebrate. *Nat. Genet.* 30:325–328.

Stöck, M., C. Steinlein, D.K. Lamatsch, M. Schartl, and M. Schmid. 2005. Multiple origins of tetraploid taxa in the Eurasian *Bufo viridis* subgroup. *Genetica* 124:255–272.

Stockard, C.R. 1921a. Developmental rate and structural expression: an experimental study of twins, "double monsters" and single deformities, and the interaction among embryonic organs during their origin and development. *Am. J. Anat.* 28:115–277.

Stockard, C.R. 1921b. A probable explanation of polyembryony in the armadillo. *Am. Nat.* 55:62–68.

Stone, R. 2006. The saola's last stand. *Science* 314:1380–1383.

Storrs, E.E., and R.J. Williams. 1968. A study of monozygous quadruplet armadillos in relation to mammalian inheritance. *Proc. Natl. Acad. Sci. USA* 60:910–914.

Strand, M.R. 1989a. Oviposition behavior and progeny allocation of the polyembryonic wasp *Copidosoma floridanum* (Hymenoptera: Encyrtidae). *J. Insect Behav.* 2:355–369.

Strand, M.R. 1989b. Clutch size, sex ratio, and mating by the polyembryonic encyrtid *Copidosoma floridanum. Fla. Entomol.* 72:32–42.

Strasburg, J.L., and M. Kearney. 2005. Phylogeography of sexual *Heteronotia binoei* (Gekkonidae) in the Australian arid zone: climatic cycling and repetitive hybridization. *Mol. Ecol.* 14:2755–2772.

Streisinger, G., C. Walker, N. Dower, D. Knauber, and F. Singer. 1981. Production of clones of homozygous diploid zebra fish (*Brachydanio rerio*). *Nature* 291:293–296.

Suomalainen, E., A. Saura, and J. Lokki. 1987. *Cytology and Evolution in Parthenogenesis.* CRC Press, Boca Raton, FL.

Tagarelli, A., A. Piro, P. Lagonia, and G. Tagarelli. 2004. Hans Spemann. One hundred years before the birth of experimental embryology. *Anat. Histol. Embryol.* 33:28–32.

Takebayashi, N., and P.L. Morrell. 2001. Is self-fertilization an evolutionary dead end? Revisiting an old hypothesis with genetic theories and macroevolutionary approach. *Am. J. Bot.* 88:1143–1150.

Tatarenkov, A., and J.C. Avise. 2007. Rapid concerted evolution in animal mitochondrial DNA. *Proc. R. Soc. Lond. B Biol. Sci.* 274:1795–1798.

Tatarenkov, A., H. Gao, M. Mackiewicz, D.S. Taylor, B.J. Turner, and J.C. Avise. 2007. Strong population structure despite evidence of recent migration in a selfing hermaphroditic vertebrate, the mangrove killifish (*Kryptolebias marmoratus*). *Mol. Ecol.* 16:2701–2711.

Taulman, J.F., and L.W. Robbins. 1996. Recent range expansion and distributional limits of the nine-banded armadillo (*Dasypus novemcinctus*) in the United States. *J. Biogeogr.* 23:635–648.

Taylor, D.S. 1990. Adaptive specializations of the cyprinodont fish, *Rivulus marmoratus*. *Fla. Sci.* 53:239–248.

Taylor, D.S. 2000. Biology and ecology of *Rivulus marmoratus:* new insights and a review. *Fla. Sci.* 63:242–255.

Taylor, D.S., M.T. Fisher, and B.J. Turner. 2001. Homozygosity and heterozygosity in three populations of *Rivulus marmoratus*. *Environ. Biol. Fishes* 61:455–459.

Thorgaard, G.H. 1983. Chromosome set manipulation and sex control in fish. Pp. 405–434 in: *Fish Physiology,* vol. IXB, W.S. Hoar, D.J. Randall, and E.M. Donaldson (eds.). Academic Press, New York.

Thorgaard, G.H., P.A. Wheeler, W.P. Young, B.D. Robison, and S.S. Ristow. 2003. Genetic analysis of complex traits in clonal rainbow trout lines. Pp. 395–398 in: *Aquatic Genomics,* N. Shimizu, T. Aoki, I. Hirono, and F. Takashima (eds.). Springer-Verlag, Tokyo.

Thornhill, N.W. 1993. *The Natural History of Inbreeding and Outbreeding.* University of Chicago Press, Chicago.

Thresher, R.E. 1984. *Reproduction in Reef Fishes.* T.F.H. Publications, Neptune City, NJ.

Tompkins, R. 1978. Triploid and gynogenetic diploid *Xenopus laevis*. *J. Exp. Zool.* 203:251–256.

Trivers, R.L. 1974. Parent-offspring conflict. *Am. Zool.* 14:249–264.

Trottier, T.M., and J.B. Armstrong. 1976. Diploid gynogenesis in the Mexican axolotl. *Genetics* 83:783–792.

Trounson, A. 2002. The genesis of embryonic stem cells. *Nat. Biotechnol.* 20:237–238.

Tunner, H.G., and H. Nopp. 1979. Heterosis in the common European water frog. *Naturwissenschaften* 66:268–269.

Turner, B.J. 1982. The evolutionary genetics of a unisexual fish, *Poecilia formosa*. Pp. 265–305 in: *Mechanisms of Speciation,* C. Barigozzi (ed.). Alan R. Liss, New York.

Turner, B.J., J.S. Balsano, P.J. Monaco, and E.M. Rasch. 1983. Clonal diversity and evolutionary dynamics in a diploid-triploid complex of unisexual fishes (*Poecilia*). *Evolution* 37:798–809.

Turner, B.J., B.L. Brett, and R.R. Miller. 1980. Interspecific hybridization and the evolutionary origin of a gynogenetic fish, *Poecilia formosa. Evolution* 343:917–922.
Turner, B.J., W.P. Davis, and D.S. Taylor. 1992a. Abundant males in populations of a selfing hermaphroditic fish—*Rivulus marmoratus*—from some Belize cays. *J. Fish Biol.* 40:307–310.
Turner, B.J., J.F. Elder Jr., T.F. Laughlin, and W.P. Davis. 1990. Genetic variation in clonal vertebrates detected by simple-sequence DNA fingerprinting. *Proc. Natl. Acad. Sci. USA* 87:5653–5657.
Turner, B.J., J.F. Elder Jr., T.F. Laughlin, W.P. Davis, and D.S. Taylor. 1992b. Extreme clonal diversity and divergence in populations of a selfing hermaphroditic fish. *Proc. Natl. Acad. Sci. USA* 89:10643–10647.
Turner, B.J., M.T. Fisher, D.S. Taylor, W.P. Davis, and B.L. Jarrett. 2006. Evolution of "maleness" and outcrossing in a population of the self-fertilizing killifish, *Kryptolebias marmoratus. Evol. Ecol. Res.* 8:1475–1486.
Uzzell, T.M. 1964. Relations of the diploid and triploid species of the *Ambystoma jeffersonianum* complex (Amphibia, Caudata). *Copeia* 1964:257–300.
Uzzell, T.M. 1969. Notes on spermatophore production by salamanders of the *Ambystoma jeffersonianum complex. Copeia* 1969:602–612.
Uzzell, T.M. 1970. Meiotic mechanisms of naturally occurring unisexual vertebrates. *Am. Nat.* 104:433–445.
Uzzell, T.M., and J.C. Barry. 1971. *Leposoma percarinatum,* a unisexual species related to *L. guianense,* and *Leposoma ioanna,* a new species from Pacific coastal Colombia. *Postilla* 154:1–39.
Uzzell, T.M., and L. Berger. 1975. Electrophoretic phenotypes of *Rana ridibunda, Rana lessonae,* and their hybridogenetic associate *Rana esculenta. Proc. Natl. Acad. Sci. USA* 127:13–24.
Uzzell, T.M., and I.S. Darevsky. 1975. Biochemical evidence for hybrid origin of the parthenogenetic species of the *Lacerta saxicola* complex (Sauria: Lacertidae), with a discussion of some ecological and evolutionary implications. *Copeia* 1975:204–222.
Uzzell, T.M., R. Günther, and L. Berger. 1977. *Rana ridibunda* and *Rana esculenta:* a leaky hybridogenetic system (Amphibia, Salientia). *Proc. Acad. Nat. Sci. Phila.* 128:147–171.
Uzzell, T., H. Hotz, and L. Berger. 1980. Genome exclusion in gametogenesis by an interspecific *Rana* hybrid: evidence from electrophoresis of individual oocytes. *J. Exp. Zool.* 214:251–259.
Van Valen, L.M. 1973. A new evolutionary law. *Evol. Theory* 1:1–30.
Vasil'ev, V.P., E.D. Vasil'eva, and A.G. Osinov. 1989. Evolution of a diploid-triploid-tetraploid complex of fishes of the genus *Cobitis* (Pisces, Cobitidae). Pp. 153–169 in: *Evolution and Ecology of Unisexual Vertebrates,* R.M. Dawley and J.P. Bogart (eds.). New York State Museum, Albany.

von Jhering, H. 1885. Ueber die Fortpflanzung dur Gurteltiere. *Sitzungsber. Akad. Wiss. Berlin Bd.* 28:567–573.

von Jhering, H. 1886. Nachtrag zur Entwicklung von *Praopus. Arch. f. Anat. u. Physiol.,* Abst.:541–542.

Vrana, K.E., J.D. Hipp, A.M. Goss, B.A. McCool, D.R. Riddle, S.J. Walker, P.J. Wettstein, L.P. Studer, V. Tabar, K. Cunniff, et al. 2003. Nonhuman primate parthenogenetic stem cells. *Proc. Natl. Acad. Sci. USA* 100:11911–11916.

Vrijenhoek, R.C. 1978. Coexistence of clones in a heterogeneous environment. *Science* 199:549–552.

Vrijenhoek, R.C. 1979. Factors affecting clonal diversity and coexistence. *Am. Zool.* 19:787–797.

Vrijenhoek, R.C. 1984a. Ecological differentiation among clones: the frozen niche-variation model. Pp. 217–231 in: *Population Biology and Evolution,* K. Wohrman and V. Loschcke (eds.). Springer-Verlag, New York.

Vrijenhoek, R.C. 1984b. The evolution of clonal diversity in *Poeciliopsis.* Pp. 399–429 in: *Evolutionary Genetics of Fishes,* B.J. Turner (ed.). Plenum Press, New York.

Vrijenhoek, R.C. 1985. Homozygosity and interstrain variation in the self-fertilizing hermaphroditic fish, *Rivulus marmoratus. J. Hered.* 76:1475–1486.

Vrijenhoek, R.C. 1989a. Genetic and ecological constraints on the origins and establishment of unisexual vertebrates. Pp. 24–31 in: *Evolution and Ecology of Unisexual Vertebrates,* R.M. Dawley and J.P. Bogart (eds.). New York State Museum, Albany.

Vrijenhoek, R.C. 1989b. Genotypic diversity and coexistence among sexual and clonal lineages of *Poeciliopsis.* Pp. 386–400 in: *Speciation and Its Consequences,* D. Otte and J.A. Endler (eds.). Sinauer, Sunderland, MA.

Vrijenhoek, R.C. 1994. Unisexual fish: model systems for studying ecology and evolution. *Annu. Rev. Ecol. Syst.* 25:71–96.

Vrijenhoek, R.C. 1998. Animal clones and diversity. *BioScience* 48:617–628.

Vrijenhoek, R.C., R.A. Angus, and R.J. Schultz. 1977. Variation and heterozygosity in sexually versus clonally reproducing populations of *Poeciliopsis. Evolution* 31:767–781.

Vrijenhoek, R.C., R.A. Angus, and R.J. Schultz. 1978. Variation and clonal structure in a unisexual fish. *Am. Nat.* 112:41–55.

Vrijenhoek, R.C., R.M. Dawley, C.J. Cole, and J.P. Bogart. 1989. A list of known unisexual vertebrates. Pp. 19–23 in: *Evolution and Ecology of Unisexual Vertebrates,* R.M. Dawley and J.P. Bogart (eds.). New York State Museum, Albany.

Vrijenhoek, R.C., and E. Pfeiler. 1997. Differential survival of sexual and asexual *Poeciliopsis* during environmental stress. *Evolution* 51:1593–1600.

Vrijenhoek, R.C., and R.J. Schultz. 1974. Evolution of a trihybrid unisexual fish (*Poeciliopsis,* Poecilidae). *Evolution* 28:306–319.

Vyas, D.K., C. Moritz, D. Peccinini-Seale, J.W. Wright, and W.M. Brown. 1990. The evolutionary history of parthenogenetic *Cnemidophorus lemniscatus* (Sauria:

Teiidae). II. Maternal origin and age inferred from mitochondrial DNA analyses. *Evolution* 44:922–932.

Wadman, M. 2007. Dolly: a decade on. *Nature* 445:800–801.

Wakayama, T., A.C. Perry, M. Zuccotti, K.R. Johnson, and R. Yanagimuchi. 1998. Full-term development of mice from enucleated oocytes injected with cumulus cell nuclei. *Nature* 394:369–374.

Walker, J.M. 1981a. Systematics of *Cnemidophorus gularis*. I. Reallocation of populations currently allocated to *Cnemidophorus gularis* and *Cnemidophorus scalaris* in Coahuila, Mexico. *Copeia* 1981:826–849.

Walker, J.M. 1981b. Systematics of *Cnemidophorus gularis*. II. Specific and subspecific identity of the Zacatecas whiptail (*Cnemidophorus gularis semiannulatus*). *Copeia* 1981:850–868.

Walker, J.M., J.E. Cordes, and M.A. Paulissen. 1989. Hybrids of two parthenogenetic clonal complexes and a gonochoristic species of *Cnemidophorus,* and the relationship of hybridization to habitat characteristics. *J. Herpetol.* 23:119–130.

Walker, J.M., J.E. Cordes, and H.L. Taylor. 1997. Parthenogenetic *Cnemidophorus tesselatus* complex (Sauria: Teiidae): a neotype for diploid *C. tesselatus* (Say, 1823), redescription of the taxon, and description of a new triploid species. *Herpetologica* 53:233–259.

Warner, R.R. 1975. The adaptive significance of sequential hermaphroditism in animals. *Am. Nat.* 109:61–82.

Warner, R.R., D.R. Robertson, and E.G. Leigh Jr. 1975. Sex change and sexual selection. *Science* 190:633–638.

Warner, R.R., and S.E. Swearer. 1991. Social control of sex change in the bluehead wrasse, *Thalassoma bifasciatum* (Pisces: Labridae). *Biol. Bull.* 181:199–204.

Watson, J.D., and F.H.C. Crick. 1953a. Molecular structure of nucleic acids: a structure for deoxyribonucleic acids. *Nature* 171:737–738.

Watson, J.D. and F.H.C. Crick. 1953b. Genetic implications of the structure of deoxyribonucleic acid. *Nature* 171:964–967.

Watts, P.C., K.R. Buley, S. Sanderson, W. Boardman, C. Ciofis, and R. Gibson. 2006. Parthenogenesis in Komodo dragons. *Nature* 444:1021–1022.

Weeks, S.C. 1995. Comparisons of life-history traits between clonal and sexual fish (*Poeciliopsis:* Poeciliidae) raised in monoculture and mixed treatments. *Evol. Ecol.* 9:258–274.

Weeks, S.C., C. Benvenuto, and S.K. Reed. 2006. When males and hermaphrodites coexist: a review of androdioecy in animals. *Integr. Comp. Biol.* 46:449–464.

Weeks, S.C., O.E. Gaggiotti, K.P. Spindler, R.E. Schenck, and R.C. Vrijenhoek. 1992. Feeding behavior in sexual and clonal strains of *Poeciliopsis*. *Behav. Ecol. Sociobiol.* 30:1–6.

Weibel, A.C., T.E. Dowling, and B.J. Turner. 1999. Evidence that an outcrossing population is a derived lineage in a hermaphroditic fish (*Rivulus marmoratus*). *Evolution* 53:1217–1225.

West, S.A., C.M. Lively, and A.F. Reed. 1999. A pluralist approach to sex and recombination. *J. Evol. Biol.* 12:1003–1012.

Westhusin, M.E., C.R. Long, T. Shin, J.R. Hill, C.R. Looney, J.H. Pryor, and J.A. Piedrahita. 2001. Cloning to reproduce desired genotypes. *Theriogenology* 55:35–49.

Wetherington, J.D., K.E. Kotora, and R.C. Vrijenhoek. 1987. A test of the spontaneous heterosis hypothesis for unisexual vertebrates. *Evolution* 41:721–731.

Wetherington, J.D., R.A. Schenck, and R.C. Vrijenhoek. 1989a. Origins and ecological success of unisexual *Poeciliopsis:* the frozen niche variation model. Pp. 259–276 in: *The Ecology and Evolution of Poeciliid Fishes,* G.A. Meffe and F.F. Snelson Jr. (eds.). Prentice Hall, Englewood Cliffs, NJ.

Wetherington, J.D., S.C. Weeks, K.E. Kotora, and R.C. Vrijenhoek. 1989b. Genotypic and environmental components of variation in growth and reproduction of fish hemiclones (*Poeciliopsis:* Poeciliidae). *Evolution* 43:635–645.

Wetzel, R.M. 1985. Taxonomy and distribution of armadillos, Dasypodidae. Pp. 23–46 in: *The Evolution and Ecology of Armadillos, Sloths, and Vermilinguas,* G.G. Montgomery (ed.). Smithsonian Institution Press, Washington, DC.

White, M.J.D. 1978. Cytogenetics of the parthenogenetic grasshopper *Warramaba* (formerly *Moraba*) *virgo* and its bisexual relatives. III. Meiosis of male "synthetic *virgo*" individuals. *Chromosoma* 67:55–61.

Whittier, J.M., D. Stewart, and L. Tolley. 1994. Ovarian and oviductal morphology of sexual and parthenogenetic geckos of the *Heteronotia binoei* complex. *Copeia* 1994:484–492.

Willadsen, S.M. 1986. Nuclear transplantation in sheep embryos. *Nature* 320:63–65.

Williams, G.C. 1957. Pleiotropy, natural selection, and the evolution of senescence. *Evolution* 11:398–411.

Williams, G.C. 1975. *Sex and Evolution.* Princeton University Press, Princeton, NJ.

Wilmut, I., A.E. Schnieke, J. McWhir, A.J. Kind, and K.H.S. Campbell. 1997. Viable offspring derived from fetal and adult mammalian cells. *Nature* 385:810–813.

Woods, G.L., K.L. White, D.K. Vanderwall, G.P. Li, K.I. Aston, T.D. Bunch, L.N. Meerdo, and B.J. Pate. 2003. A mule cloned from fetal cells by nuclear transfer. *Science* 301:1063.

Wright, J.W., and C.H. Lowe. 1968. Weeds, polyploids, parthenogenesis and the geographical and ecological distribution of all-female species of *Cnemidophorus. Copeia* 1968:128–138.

Wright, J.W., C. Spolsky, and W.M. Brown. 1983. The origin of the parthenogenetic lizard *Cnemidophorus laredoensis* inferred from mitochondrial DNA analysis. *Herpetologica* 39:410–416.

Wright, J.W., and L.J. Vitt, eds. 1993. *Biology of Whiptail Lizards: Genus Cnemidophorus.* Oklahoma Museum of Natural History, Norman.

Wyatt, R. 1988. Phylogenetic aspects of the evolution of self-pollination. Pp. 109–131 in: *Plant Evolutionary Biology,* L.D. Gottlieb and S.K. Jain (eds.). Chapman and Hall, London.

Wynn, A.L., C.J. Cole, and A.L. Gardner. 1987. Apparent triploidy in the unisexual brahminy blind snake *Rhamphotyphlops braminus. Am. Mus. Novit.* 2868:1–7.

Yang, L., S.-T. Yang, X.-H. Wei, and J.-F. Gui. 2001. Genetic diversity among different clones of the gynogenetic silver crucian carp, *Carassius auratus gibelio,* revealed by transferrin and isozyme markers. *Biochem. Genet.* 39:213–225.

Yonenaga-Yassuda, Y., P.E. Vanzolini, M.T. Rodrigues, and C.M. de Carvalho. 1995. Chromosome banding patterns in the unisexual microteiid *Gymnophthalmus underwoodi* and in two related sibling species (Gymnophthalmidae, Sauria). *Cytogenet. Cell Genet.* 70:29–34.

Young, W.P., P.A. Wheeler, V.H. Coryell, P. Keim, and G.H. Thorgaard. 1998. A detailed genetic linkage map of rainbow trout produced using doubled haploids. *Genetics* 148:839–850.

Young, W.P., P.A. Wheeler, R.D. Fields, and G.H. Thorgaard. 1996. DNA fingerprinting confirms isogenicity of androgenetically derived rainbow trout lines. *J. Hered.* 87:77–81.

Zeng, R., and Y.-Z. Zeng. 2005. Molecular cloning and characterization of *SLA-DR* genes in the 133-family of the Banna mini-pig inbred line. *Anim. Genet.* 36:258–286.

Zhou, L., Y. Wang, and J.F. Gui. 2000a. Analysis of genetic heterogeneity among five gynogenetic clones of silver crucian carp, *Carassius auratus gibelio* Bloch, based on detection of RAPD molecular markers. *Cytogenet. Cell Genet.* 88:131–139.

Zhou, L., Y. Wang, and J.F. Gui. 2000b. Genetic evidence for gonochoristic reproduction in gynogenetic silver Crucian carp (*Carassius auratus gibelio* Bloch) as revealed by RAPD assays. *J. Mol. Evol.* 51:498–506.

Zhou, Q.I., J.P. Renard, G. Le Friec, V. Brochard, N. Beaujean, Y. Cherifi, A. Fraichard, and J. Cozzi. 2003. Generation of fertile cloned rats by regulating oocyte activation. *Science* 302:1179.

Zinchenko, V.L., and M.V. Ivashin. 1987. Polyembryony and developmental abnormalities in minke whales (Baleanoptera, Acutorostrata) of the Southern Hemisphere. *Zool. J.* 66:1975–1976.

Zug, G.R. 1991. *The Lizards of Fiji: Natural History and Systematics.* Bernice Bishop Museum Press, Honolulu, HI.

INDEX